JOURNAL OF APPLIED LOGICS - IFCOLOG
JOURNAL OF LOGICS AND THEIR APPLICATIONS

Volume 5, Number 2

April 2018

Disclaimer

Statements of fact and opinion in the articles in Journal of Applied Logics - IfCoLog Journal of Logics and their Applications (JAL-FLAP) are those of the respective authors and contributors and not of the JAL-FLAP. Neither College Publications nor the JAL-FLAP make any representation, express or implied, in respect of the accuracy of the material in this journal and cannot accept any legal responsibility or liability for any errors or omissions that may be made. The reader should make his/her own evaluation as to the appropriateness or otherwise of any experimental technique described.

ISBN 978-1-84890-277-0
ISSN (E) 2055-3714
ISSN (P) 2055-3706

College Publications
Scientific Director: Dov Gabbay
Managing Director: Jane Spurr

http://www.collegepublications.co.uk

Printed by Lightning Source, Milton Keynes, UK

Editorial Board

Modal and Temporal Logic
Carlos Areces
Melvin Fitting
Victor Marek
Mark Reynolds.
Frank Wolter
Michael Zakharyaschev

Automated Inference Systems and Model Checking
Ed Clarke
Ulrich Furbach
Hans Juergen Ohlbach
Volker Sorge
Andrei Voronkov
Toby Walsh

Formal Methods: Specification and Verification
Howard Barringer
David Basin
Dines Bjorner
Kokichi Futatsugi
Yuri Gurevich

Logic and Software Engineering
Manfred Broy
John Fitzgerald
Kung-Kiu Lau
Tom Maibaum
German Puebla

Logic and Constraint Logic Programming
Manuel Hermenegildo
Antonis Kakas
Francesca Rossi
Gert Smolka

Logic and Databases
Jan Chomicki
Enrico Franconi
Georg Gottlob
Leonid Libkin
Franz Wotawa

Logic and Physics (space time. relativity and quantum theory)
Hajnal Andreka
Kurt Engesser
Daniel Lehmann
lstvan Nemeti
Victor Pambuccian

Logic for Knowledge Representation and the Semantic Web
Franz Baader
Anthony Cohn
Pat Hayes
Ian Horrocks
Maurizio Lenzerini
Bernhard Nebel

Tactical Theorem Proving and Proof Planning
Alan Bundy
Amy Felty
Jacques Fleuriot
Dieter Hutter
Manfred Kerber
Christoph Kreitz

Logic and Algebraic Programming
Jan Bergstra
John Tucker

Logic in Mechanical and Electrical Engineering
Rudolf Kruse
Ebrahaim Mamdani

Logic and Law
Jose Carmo
Lars Lindahl
Marek Sergot

Applied Non-classical Logic
Luis Farinas del Cerro
Nicola Olivetti

Mathematical Logic
Wilfrid Hodges
Janos Makowsky

Cognitive Robotics: Actions and Causation
Gerhard Lakemeyer
Michael Thielscher

Type Theory for Theorem Proving Systems
Peter Andrews
Chris Benzmüller
Chad Brown
Dale Miller
Carsten Schlirmann

Logic Applied in Mathematics (including e-Learning Tools for Mathematics and Logic)
Bruno Buchberger
Fairouz Kamareddine
Michael Kohlhase

Logic and Computational Models of Scientific Reasoning
Lorenzo Magnani
Luis Moniz Pereira
Paul Thagard

Logic and Multi-Agent Systems
Michael Fisher
Nick Jennings
Mike Wooldridge

Logic and Neural Networks
Artur d'Avila Garcez
Steffen Holldobler
John G. Taylor

Logic and Planning
Susanne Biundo
Patrick Doherty
Henry Kautz
Paolo Traverso

Algebraic Methods in Logic
Miklos Ferenczi
Rob Goldblatt
Robin Hirsch
Idiko Sain

Non-monotonic Logics and Logics of Change
Jurgen Dix
Vladimir Lifschitz
Donald Nute
David Pearce

Logic and Learning
Luc de Raedt
John Lloyd
Steven Muggleton

Logic and Natural Language Processing
Wojciech Buszkowski
Hans Kamp
Marcus Kracht
Johanna Moore
Michael Moortgat
Manfred Pinkal
Hans Uszkoreit

Fuzzy Logic Uncertainty and Probability
Didier Dubois
Petr Hajek
Jeff Paris
Henri Prade
George Metcalfe
Jon Williamson

SCOPE AND SUBMISSIONS

This journal considers submission in all areas of pure and applied logic, including:

pure logical systems
proof theory
constructive logic
categorical logic
modal and temporal logic
model theory
recursion theory
type theory
nominal theory
nonclassical logics
nonmonotonic logic
numerical and uncertainty reasoning
logic and AI
foundations of logic programming
belief revision
systems of knowledge and belief
logics and semantics of programming
specification and verification
agent theory
databases

dynamic logic
quantum logic
algebraic logic
logic and cognition
probabilistic logic
logic and networks
neuro-logical systems
complexity
argumentation theory
logic and computation
logic and language
logic engineering
knowledge-based systems
automated reasoning
knowledge representation
logic in hardware and VLSI
natural language
concurrent computation
planning

This journal will also consider papers on the application of logic in other subject areas: philosophy, cognitive science, physics etc. provided they have some formal content.

Submissions should be sent to Jane Spurr (jane.spurr@kcl.ac.uk) as a pdf file, preferably compiled in LaTeX using the IFCoLog class file.

CONTENTS

ARTICLES

EDITORIAL

SERENA VILLATA
Université Côte d'Azur, CNRS, Inria, I3S, France
villata@i3s.unice.fr

This special issue contains the journal version of four contributions to the Handbook of Normative Multi-Agent Systems (NorMAS), which will appear at College Publications. The NorMAS initiative aims at providing a comprehensive coverage of both the state of the art and future research perspectives in the interdisciplinary field of normative multi-agent systems. It is meant to be an open community effort and a service to current and future students and researchers interested in this field. We invite the readers to buy the forthcoming handbook for a full view. Please visit the website for more information and feel free to send us comments, suggestions and proposals: http://normativemas.org/

The articles in this special issue and the chapters in the handbook give a survey of the area and may also contain a more personal view. For the survey part, at least the work reported in the NorMAS conference series is discussed. Instead of just a historical overview, the authors also address new developments, open topics and emerging areas. The handbooks appeal to all disciplines, including logic, computer science, law, philosophy, and linguistics. The articles in this special issue discuss the obtained results and open challenges in modeling normative multiagent systems during the last two decades. More information can be found in the handbook on deontic logic and normative systems, or the website http://deonticlogic.org/

The first paper in this special issue formally analyses the issue of modeling norm specification and verification in multiagent systems (MAS). In particular, for Alechina *et al.*, violation conditions of regulative norms may correspond to conditions on states, actions, or arbitrary temporal patterns. They may be specified semantically or expressed syntactically in a suitable temporal logic, or in a programming language. Verification problems for norms or rather for normative systems involve verifying consistency of norms, verifying whether violation conditions hold, and finally verifying whether a system where norms are enforced satisfies some system objective.

Vol. 5 No. 2 2018
Journal of Applied Logics — IFCoLog Journal of Logics and their Applications

In the second paper, Pigozzi and Frantz discuss how to model norm dynamics in multiagent systems. More precisely, they review of all existing life cycle models looking at normative processes from a holistic perspective, which include the introduction of individual life cycle models and their contextualization with specific contributions that exemplify life cycle processes. They provide a comprehensive contemporary overview of individual contributions to the area of NorMAS and the systematic comparison of the discussed life cycle models. Based on this analysis, they also propose a refined life cycle model that resolves terminological ambiguities and ontological inconsistencies of the existing models, while reflecting the contemporary view on norm formation and emergence.

In the third chapter, Fornara and Balke analyse possible solutions to model organizations and institutions in multiagent systems. Institutions and organizations are two concepts within the MAS community that are commonly referred to when the question arises on how to ensure that an (open) MAS exhibits some desired properties, while the agents interacting in that MAS have some degree of autonomy at the same time. The authors give a brief introduction to the two concepts as its related ideas, outlining research done in the area of NorMAS and giving pointers on current challenges for modeling institutions and organizations.

The fourth chapter aims at discussing research directions towards the modeling of those norms that are embedded in the society, with a particular attention to ethics and sensitive design. After elaborating on the notions of decision rights, responsibility and accountability, Christiaanse ends up with rephrasing the original design question into seven key questions formulated in a principled way using the procedure to classify and analyze an ethical system applied to the code of conduct of Nike. Also the issue of defining the notion of a model as the start and result of the design process is addressed in this chapter.

Received 13 April 2018

NORM SPECIFICATION AND VERIFICATION IN MULTI-AGENT SYSTEMS

NATASHA ALECHINA
University of Nottingham, Nottingham, NG8 1BB UK
natasha.alechina@nottingham.ac.uk

MEHDI DASTANI
Utrecht University, Princetonplein 5, 3584 CC Utrecht, The Netherlands
m.m.dastani@uu.nl

BRIAN LOGAN
University of Nottingham, Nottingham, NG8 1BB UK
brian.logan@nottingham.ac.uk

Abstract

This article presents a high-level overview of the literature on norms and their uses in multi-agent systems. We distinguish the main types of norms used in multi-agent systems, and the ways in which the behaviour of a system can be modified through the enforcement of norms. We first review the formal approaches used to study norms and norm enforcement mechanisms. We then explain the syntax and semantics of the key specification languages used to represent norms, and briefly survey some programming frameworks that support the implementation of normative multi-agent systems. Finally, we briefly review the key research questions and techniques in the important area of norm verification.

1 Introduction

Norms are generally conceived as standards of behaviour [19, 33]. In the norm literature (e.g., [19, 20, 33]), various norm types have been distinguished based on the authorities that issue and enforce the norms. Examples of norm types are legal, social, moral and rational norms. Legal norms requires a legal body that issues norms and a corresponding executive body that enforces norms. For example, the legislative body of a state can issue traffic laws and the executive body of the state can enforce the traffic laws. Social norms often emerge through interaction within a community of individuals, who are subsequently in charge of enforcing the norm. For example, the amount of labour in a workplace can emerge as a social norm, after which those who work too hard or too little get criticised or even ignored/excluded from the workplace. Moral norms differ from legal and social norms as there is no authority or society required to issue and enforce moral norms. Moral norms are seen as a product of reasoning or internalisation of some external standards. Individuals follow their moral norms because of other internal reasons such as deliberation or emotions. Finally, rational norms includes prescriptive rational rules such as axioms of logics or equilibria in games. In general, norms are prescriptive in the sense that they prescribe which states, actions or behaviour to pursue or avoid.

In multi-agent systems, norms are often used to ensure the overall objectives of the system. In order to organise a multi-agent system in such a way that the standards of behaviour are actually followed by the agents, norms should be enforced by means of regimentation or sanctioning, e.g., [50, 43, 31]. When regimenting norms, agents' behaviours leading to violations of norms are made impossible. Regimentation prevents agents from reaching a forbidden state or performing a forbidden action. Enforcing norms by regimentation decreases agent autonomy significantly. Norms can be regimented in various ways. For example, norms can be incorporated in the agent's decision making mechanisms so that all the agent's executions will be compliant with the norms. Norms can also be regimented externally by ignoring violating actions or undoing their effects. In the latter case, the enforcement mechanism is assumed to have control over the effects of the agents' actions, e.g., the enforcement mechanism can decide not to pass messages between some agents or to undo the effect of the agents' actions in the multi-agent environment. Instead, norms enforcement can be based on the idea of responding after a violation of the norms has occurred. Such a response, which includes sanctions, aims to return the system to an optimal state. For sanction-based enforcement it is essential that the norm violating actions are observable by the system (e.g., fines can be issued in traffic systems only if the speed of cars can be observed). Sanction-based enforcement allows agents to violate norms and therefore contributes to the flexibility and autonomy of the agents' behaviour [26].

One of the key questions regarding norm enforcement in multi-agent systems is whether the enforcement of a given set of norms can ensure some given desirable system properties. In particular, provided that a multi-agent system does not satisfy some given desirable system properties, does the enforcement of a given set of norms modify the system in such a way that the desirable system properties are ensured. This problem is one of the versions of norm verification problem. Another related problem is to generate a set of norms that, when enforced in the system, ensures the desirable system properties. This latter problem is called norm synthesis problem. Both problems require a procedure to update a system with a set of norms. Such a procedure implements a norm enforcement mechanism. Another key question regarding norm enforcement is the expressive power of norms. In general, there is a trade-off between expressiveness of norms and the computational complexity of the verification, synthesis and update problems: the more expressive norms, the higher computational complexity of the problems.

In this chapter, we ignore the problem of norm synthesis and cover approaches to specification and verification of normative systems related to regulative norms, that is norms that can be violated. We survey various approaches to norm specification and cover different types of regulative norms such as state-, action-, and behaviour-based norms[1]. For verification, we only cover approaches using model-checking, because they are by far the more prevalent. However, there exists work using theorem proving for verification, for example [42].

2 Background

behaviour-based In this section we introduce the necessary background on transition systems and temporal logics used in the specification and verification of norms. This includes background on temporal logics such as Linear-Time Temporal Logic LTL [58], Computation Tree Logic CTL and CTL^* [28], Alternating-Time Temporal Logic ATL and ATL^* [11]. In the exposition of LTL, LTL + Past, CTL, and CTL^* below, we largely follow the notation in [60].

The logical languages we introduce below are defined relative to a set of propositional atoms Π, and talk about *state transition systems*, or transition systems for brevity. A transition system is a graph where states are vertices (decorated with propositional atoms) and transitions are edges. In a *labelled* transition system, edges are also decorated with labels, or action names.

Definition 1 (State Transition System). *A state transition system is a tuple $M = (S, R, V)$, where S is a finite, non-empty set of states, $R \subseteq S \times S$ is a transition relation (for simplicity, we assume that R is a total relation, that is, some transition is possible in every state), and V is a propositional valuation $S \longrightarrow 2^{\Pi}$. A pointed transition system is a pair (M, s_I), where M is a transition system and $s_I \in S$ is the initial state. A labelled transition system is built using a set L of labels. It is a tuple $M = (S, \{R_a : a \in L\}, V)$, where each $R_a \subseteq S \times S$ is a transition relation.*

A transition system can be used to describe the lifecycle of an agent, or a business process, or a system consisting of multiple interacting processes or agents. States correspond to configurations of the system at a moment in time. The transition relation corresponds to

---

[1]Action-based norms is the term most widely used in the literature; sometimes we refer to those norms as *transition-based* to cover both norms specified in terms of actions and in terms of events. We will also sometimes refer to norms specified in terms of behaviours or temporal patterns as *path-based* norms.

actions or events which change the state, and the valuation function assigns a set of atoms to a state (intuitively, the set of atoms which hold in that state).

Given a state transition system $M = (S, R, V)$, a *path* through M is a sequence $s_0, s_1, s_3, \ldots$ of states such that $s_i R s_{i+1}$ for $i = 0, 1, \ldots$. A *fullpath* is a maximal path and a *run* of M is a fullpath which starts from a state $s_I \in S$ designated as the initial state of M. We denote runs by $\rho, \rho', \ldots$, and the state at position i on ρ by $\rho[i]$. Intuitively, a path represents a finite history of events in the system, and a run corresponds to a complete infinite history or computation of the system. We denote the set of all runs in M by $\mathcal{P}(M)$.

For a state $s \in S$, the *tree* rooted at s is the infinite tree $T(s)$, obtained by unfolding M from s (the nodes of T are finite paths starting from s ordered by the prefix relation). $T(M) = T(s_I)$ is the *computation tree* of M. Note that branches of $T(M)$ are runs of M.

Linear Time Temporal Logic (LTL) The syntax of LTL is defined as follows:

$$\phi, \psi ::= p \mid \neg\phi \mid \phi \wedge \psi \mid \mathcal{X}\phi \mid \phi\mathcal{U}\psi$$

where $p \in \Pi$, $\neg$ stands for not, $\wedge$ for and, $\mathcal{X}$ means next state, and $\mathcal{U}$ stands for until. Other propositional connectives $\vee$ (or) and $\rightarrow$ (implies) are defined in a standard way. It is also possible to define $\mathcal{F}\phi$ (ϕ holds some time in the future) as $\top\mathcal{U}\phi$ and $\mathcal{G}\phi$ (always ϕ) as $\neg\mathcal{F}\neg\phi$.

The truth definition for formulas of LTL is given inductively with respect to a run $\rho \in \mathcal{P}(M)$ and a position i on ρ. We omit M, $\mathcal{P}(M)$, $T(M)$ etc. when it is clear from the context:

$\rho, i \models p$ iff $p \in V(\rho[i])$

$\rho, i \models \neg\phi$ iff $\rho, i \not\models \phi$

$\rho, i \models \phi \wedge \psi$ iff $\rho, i \models \phi$ and $\rho, i \models \psi$

$\rho, i \models \mathcal{X}\phi$ iff $\rho, i + 1 \models \phi$

$\rho, i \models \phi\mathcal{U}\psi$ iff $\exists j \geq i$ such that $\rho, j \models \psi$ and $\forall k : i \leq k < j, \rho, k \models \phi$

A run ρ satisfies an LTL formula ϕ if $\rho, 0 \models \phi$. A transition system M satisfies an LTL formula ϕ, written as $M \models \phi$, if all runs in $\mathcal{P}(M)$ satisfy ϕ.

Extending LTL with Path Quantifiers The syntax of CTL^* is defined as follows:

$$\phi, \psi ::= p \mid \neg\phi \mid \phi \wedge \psi \mid \mathcal{X}\phi \mid \phi\mathcal{U}\psi \mid E\phi$$

(adding a quantifier over paths E, with the intended meaning 'there exists a continuation of the run satisfying ϕ). The universal quantifier $A\phi$ (on all runs) is defined as $\neg E\neg\phi$.

The truth definition for LTL is extended with

$\rho, i \models E\phi$ iff for some run $\rho' \in T(M)$ which is identical to ρ on the first i indices, $\rho', i \models \phi$.

A CTL^* formula ϕ is true in a transition system M, $M \models \phi$, iff $\rho, 0 \models \phi$ for all runs ρ in $T(M)$.

Note that any LTL formula is a CTL^* formula. A system M satisfies an LTL formula ϕ iff it satisfies a CTL^* formula $A\phi$. CTL^* is strictly more expressive than LTL. For example, it can express the existence of a choice point: there is a future where in the next state p holds, and a future where in the next state $\neg p$ holds, $E\mathcal{X}p \wedge E\mathcal{X}\neg p$.

Computation Tree Logic CTL is the fragment of CTL^* where every temporal modality ($\mathcal{U}$ or $\mathcal{X}$) must be under the immediate scope of a path quantifier (E or A). The semantics is inherited from CTL^*. Alternatively, the logic can be defined as follows, independently from CTL^*. The syntax is

$$\phi, \psi ::= p \mid \neg\phi \mid \phi \wedge \psi \mid E\mathcal{X}\phi \mid E(\phi\mathcal{U}\psi) \mid A(\phi\mathcal{U}\psi)$$

The semantics can be defined without reference to runs, only to states corresponding to positions on a run, as follows:

$s \models E\mathcal{X}\phi$ iff there is a branch of the tree $T(M)$ starting from s such that for the next state s' on that branch, $s' \models \phi$

$s \models E(\phi\mathcal{U}\psi)$ iff there is a branch ρ of the tree $T(M)$ with $\rho[i] = s$ such that there exists a state $s_j = \rho[j]$, $j \geq i$, on that branch such that $s_j \models \psi$ and for all states $s_k = \rho[k]$ with $i \leq k < j$, $s_k \models \phi$

$s \models A(\phi\mathcal{U}\psi)$ iff for all branches ρ of the tree $T(M)$ with $\rho[i] = s$ there exists a state $s_j = \rho[j]$, $j \geq i$, on that branch such that $s_j \models \psi$ and for all states $s_k = \rho[k]$ with $i \leq k < j$, $s_k \models \phi$

Linear Time Temporal logic with Past   Although the expressive power of temporal logics does not change with the addition of past operators, it is convenient to consider temporal logics which talk not just about the future, but also about the past.

The syntax of $LTL + Past$ formulas is defined as follows:

$$p \in \Pi \mid \neg\phi \mid \phi \wedge \psi \mid \mathcal{X}\phi \mid \phi\mathcal{U}\psi \mid \mathcal{X}^{-1}\phi \mid \phi\mathcal{S}\psi$$

where $\mathcal{X}^{-1}$ means previous state, $\mathcal{S}$ stands for since (as in, ϕ has been true since ψ became true). The truth definition for formulas is given relative to $T(M)$, a run ρ and the state at position i on ρ:

$\rho, i \models \mathcal{X}^{-1}\phi$ iff $i > 0$ and $\rho, i - 1 \models \phi$

$\rho, i \models \phi \mathcal{S} \psi$ iff $\exists j \leq i$ such that $\rho, j \models \psi$ and $\forall k : i \geq k > j, \rho, s_k \models \phi$

Alternating Time Temporal Logic (ATL) ATL formulas are interpreted on concurrent game structures.

Definition 2 (Concurrent Game Structure). *A Concurrent Game Structure (CGS) is a tuple $M = (S, V, a, \delta)$ which is defined relative to a set of agents $\mathcal{A} = \{1, \ldots, n\}$ and a set of propositional variables Π, where:*

- *S is a non-empty set of states*

- *$V : S \rightarrow \wp(\Pi)$ is a function which assigns each state in S a subset of propositional variables*

- *$a : S \times \mathcal{A} \rightarrow \mathbb{N}$ is a function which indicates the number of available moves (actions) for each player $i \in \mathcal{A}$ at a state $s \in S$ such that $a(s, i) \geq 1$. At each state $s \in S$, we denote the set of joint moves available for all players in $\mathcal{A}$ by $A(s)$. That is*

$$A(s) = \{1, \ldots, a(s, 1)\} \times \ldots \times \{1, \ldots, a(s, n)\}$$

- *$\delta : S \times \mathbb{N}^{|\mathcal{A}|} \rightarrow S$ is a partial function where $\delta(s, m)$ is the next state from s if the players execute the move $m \in A(s)$.*

The language of ATL is defined as follows:

$$p \in \Pi \mid \neg\phi \mid \phi \wedge \psi \mid \langle\langle C \rangle\rangle \mathcal{X}\phi \mid \langle\langle C \rangle\rangle G\phi \mid \langle\langle C \rangle\rangle \phi \mathcal{U} \psi$$

where $C \subseteq \mathcal{A}$. Intuitively, $\langle\langle C \rangle\rangle \gamma$ means 'the group of agents C has a strategy, all executions of which satisfy the formula γ, whatever the other agents in $\mathcal{A} \setminus C$ do'.

Definition 3 (Move). *Given a CGS M and a state $s \in S$, a move (or joint action) for a coalition $C \subseteq \mathcal{A}$ is a tuple $\sigma_C = (\sigma_i)_{i \in C}$ such that $1 \leq \sigma_i \leq a(s, i)$.*

By $A_C(s)$ we denote the set of all moves for C at state s. Given a move $m \in A(s)$, we denote by m_C the actions executed by C, $m_C = (m_i)_{i \in C}$. The set of all possible outcomes of a move $\sigma_C \in A_C(s)$ at state s is defined as follows:

$$out(s, \sigma_C) = \{s' \in S \mid \exists m \in A(s) : m_C = \sigma_C \wedge s' = \delta(s, m)\}$$

Definition 4 (Strategy). *Given a CGS M, a strategy for a subset of players $C \subseteq \mathcal{A}$ is a mapping F_C which associates each finite path $s_I, \ldots, s$ to a move in $A_C(s)$.*

A fullpath ρ is consistent with F_C iff for all $i \geq 0$, $\rho[i+1] \in out(\rho[i], F_C(\rho[0], \ldots, \rho[i]))$. We denote by $out(s, F_C)$ the set of all such fullpaths ρ starting from s, i.e. where $\rho[0] = s$. The truth definition, as for CTL, can be given relative to states in M:

- $s \models \langle\!\langle C \rangle\!\rangle \mathcal{X}\phi$ iff there exists a strategy F_C which such that for all $\rho \in out(s, F_C)$, $\rho[1] \models \phi$

- $s \models \langle\!\langle C \rangle\!\rangle \mathcal{G}\phi$ iff there exists a strategy F_C such that for all $\rho \in out(s, F_C)$, $\rho[i] \models \phi$ for all $i \geq 0$

- $s \models \langle\!\langle C \rangle\!\rangle \phi \mathcal{U}\psi$ iff there exists a strategy F_C such that for all $\rho \in out(s, F_C)$, there exists $i \geq 0$ such that $\rho[i] \models \psi$ and $\rho[j] \models \phi$ for all $j \in \{0, \ldots, i-1\}$

3 Norm Specification

In this section, we discuss how norms can be stated precisely and what it means for a norm to be violated. Many different approaches to specifying norms can be found in the literature. For example, some authors specify norms semantically, with respect to some formal model of the system (e.g., given a specification of the system, we can state that certain actions which are possible under this specification are forbidden by a norm), while others specify norms syntactically, as expressions of a formal language.[2] Alternatively, norms may be specified directly in terms of programming constructs. The specification may also depend on how the norms are enforced (regimentation or sanctioning), whether the subject of the norm is a single agent or a group of agents, etc. We therefore base our classification on whether a particular approach to norms specifies norms and their violation in terms of states (Section 3.1), in terms of actions or transitions (Section 3.2), or in terms of paths or behaviours (Section 3.3). We show how norms specified in terms of transitions and paths can be (re)expressed in temporal logic, allowing different approaches to specifying norms (and their violation) to be precisely compared. In Section 3.4 we address recent arguments that 'real life' norms cannot be expressed in temporal logics. Finally, in Section 3.5, we classify proposals for norm programming frameworks in the literature in terms of whether they can express state, transition or behaviour norms.

---

[2]The distinction is often somewhat blurred, as specification of the system is also usually done in some formal language.

3.1 State-based Norms

Norms can be specified in terms of (properties of) states. For example, in [3, 31] norms are specified by means of a set of violating states (the set of norm compliant states is the complement of the set of violating states). A state-based norm may prohibit or require states, e.g., a car is prohibited to park at a certain location or a car is obliged to have insurance. State-based norms may apply to a single agent or to a group of agents, for example, it may be prohibited for more than eight people to be in an elevator at the same time. It is generally assumed that norms may conflict, e.g., a car may be prohibited to park at a certain location while it is obliged to load a cargo at the same location. In order to specify state-based norms, various proposals have been put forward. In the rest of this section, we survey some of these proposals.

3.1.1 Counts-as Norms

Specifying norms directly in terms of states can sometimes be cumbersome. 'Counts-as' rules allow specification of norms in terms of properties (sets of states).

Regulative norms, also called deontic norms, can be seen as statements classifying system's states as complying or violating. Counts-as rules, together with a specific 'violation' atom $Viol$, can be used to classify system states. The so-called "counts-as reduction" of deontic norms builds on the tradition of the reductionistic approach in deontic logic started with the work of Anderson [12, 14, 13] and Kanger [51]. The idea of such reductionist approach is that the statement "ϕ is obligatory" in interpreted as the statement "$\neg\phi$ necessarily implies a violation" (i.e., $\neg\phi$ is prohibited), represented by the counts-as rule $\neg\phi \Rightarrow Viol$. Conditional deontic norms of the form "if C, ϕ is prohibited" can be represented by counts-as rules of the forms "ϕ counts-as $Viol$ in context C".

In contrast to regulative norms, constitutive norms establish a social institution by creating and classifying new facts, called institutional facts [61]. Institutional facts build on brute and institutional facts, and define new institutional facts. Following Searle [61], constitutive norms create and classify institutional facts by statements of the form "ϕ counts as ψ in context C", where ϕ is a brute or institutional fact and ψ is an institutional fact. In this way, a constitutive norm can be seen as defining institutional facts. Counts-as rules are used to represent constitutive norms [21, 5]. Note that the $Viol$ atom used in regulative norms can be seen as an institutional fact with the special interpretation indicating that some facts are considered as violating states.

Representing deontic norms using counts-as statements, one can consider ϕ as denoting *brute* facts (system's states), while the $Viol$ atom denotes an institutional fact. In this way, norms impose *institutional* descriptions upon the brute ones, e.g. "ϕ is a violation state". In the case of constitutive norms, one can consider ϕ as denoting brute or institutional fact, while ψ denotes institutional facts. Thus, a constitutive norm defines which brute or institutional fact can be considered as institutional fact. Counts-as statements could be complex and exhibit rich logical structure as shown, for instance, in [43]. Counts-as rules are often used with an additional context condition that specifies the applicability of the counts-as rule. For example, $\phi \Rightarrow Viol$ in ψ indicates that in the context denoted by ψ, the brute fact ϕ is considered as a violation and thus prohibited. In [21], regulative and constitutive norms are modelled as agents' goals and beliefs, respectively, which are in turn specified by rules in input/output logic. They show how counts-as relations, which represent norms, can formally be specified in input/output logic.

Counts-as rules are also used to represent norms and sanctions in an organisational setting. For example, in [31] counts-as rules are used to specify a normative artefact that is responsible for the control and coordination of software agents, and in [24] counts-as rules are used to determine a game theoretic mechanism that enforces certain socially preferred outcomes. To specify regimenting and sanctioning norms in normative artefact, special violation atoms $\texttt{viol}_\perp$ and $\texttt{viol}_i$ are introduced, respectively. These special atoms constitute the consequent of counts-as rules to represent obligations and prohibitions. For example, the counts-as rule { $\texttt{book(a)}$, $\texttt{late(a)}$ } $\Rightarrow$ {$\texttt{viol}_1$} represents the library norm which states that it is forbidden to being late in returning book $\texttt{a}$. Note that this prohibition is a sanctioning norm as it uses $\texttt{viol}_1$ atom (instead of $\texttt{viol}_\perp$). For each violation atom $\texttt{viol}_i$ a counts-as rule can be used to represent how to sanction such a violation. For example, the counts-as rule $\{viol_1\} \Rightarrow \{fined\}$ indicates that a sanctioning fine should incur in response to the violation $\texttt{viol}_1$. A normative artefact controls and coordinates the agents' activities by determining the effect of the agents' actions in their environment. An artefact is assumed to observe the agents' actions, to evaluate them with respect to a given set of norms, and to determine the effects of these actions. The realising effects can be ignoring the action effect in case a regimenting norm is violated, or adding sanctions to the resulting states in case a sanctioning norm is violated. The decisions as to which norms are violated and which sanctions should be imposed are determined by taking the closure of the environment state, where the agents' actions are performed, under the sets of counts-as rules representing norms and sanctions.

3.1.2 Norms as Defeasible Rules

Norms are often conflicting and require formalisms to capture and cope with conflicts. One possible formalism to represent conflicting norms is by means of defeasible rules. In BOID [22], defeasible rules are used to represent an agent's mental and motivational attitudes. In this framework, norms are considered as constituting an agent's motivational attitude that is used in the agent's deliberation process to determine the agent's behaviour. An agent can have conflicting motivational attitudes, e.g., an agent's obligation may conflict with other obligations or even with the agent's desires or intentions. In BOID, mental and motivational attitudes are represented by defeasible rules of the form $a \overset{x}{\hookrightarrow} b$, where $x \in \{B, O, I, D\}$ denotes possible mental attitudes such as beliefs, obligations, intentions and desires. A rule of the form $a \overset{O}{\hookrightarrow} b$ is interpreted as "if a is derived as a goal, then the agent is obliged that b is a goal". The goal generation operation, within the BOID deliberation process, applies defeasible rules iteratively and on the basis of a given order on rules to derive maximally consistent set of goals. It should be noted that norms in BOID are restricted to obligations, which are considered as a motivational attitude of an agent. Obligations in BOID are propositional properties (certain states are obligatory), similar to beliefs, desires, and intentions. Violation conditions of obligations in BOID can therefore be expressed in propositional logic. Since agents are allowed to have conflicting obligations and some obligations are not included in the set of maximally consist set of goals, an agent may comply with some and violate other norms. The following set of defeasible rules specifies an example of a BOID agent, who intends to attend a conference, is obliged to have a cheap room close to a conference site, but believes there are no cheap hotels nearby the conference site:

cheap_room $\overset{B}{\hookrightarrow}$ ¬close_to_conf_site

close_to_conf_site $\overset{B}{\hookrightarrow}$ ¬cheap_room

$\top \overset{I}{\hookrightarrow}$ go_to_conference

go_to_conference $\overset{O}{\hookrightarrow}$ cheap_room

go_to_conference $\overset{O}{\hookrightarrow}$ close_to_conf_site

3.2 Action-based Norms

As with state-based approaches to specifying norms, norms specified in terms of transitions (e.g., actions, events), can be specified directly as a set of (prohibited) transitions [1, 2, 52] (with compliant transitions defined as the complement of the set of violating transitions). Norms specified in terms of transitions may apply to an action by a single agent or an action performed by a group of agents, for example, it may be prohibited that more than 3 school children enter a shop together [4].

Action or event-based norms are used in frameworks for specifying *institutions* or agent societies, see e.g., [29].

3.3 Behaviour-based Norms

As with norms specified in terms of states and transitions, norms specified in terms of paths or temporal patterns of behaviour can be specified directly as the set of violating runs (with compliant runs defined as the complement of the violating runs) [6, 25]. However, when the number of traces is infinite, alternative approaches are necessary. As with state and action-based approaches, norms specified in terms of behaviours may apply to a single agent or to a group of agents.

3.3.1 Conditional Norms with Deadlines and Sanctions

Conditional norms with deadlines and sanctions were introduced in [32]. Conditional norms are triggered (detached) in certain states of the environment and have a temporal dimension specified by a deadline. The satisfaction or violation of a detached norm depends on whether the behaviour of the agent(s) brings about a specified state of the environment before a state in which the deadline condition is true. Norms can be enforced by means of sanctions or they can be regimented by disabling actions in specific states.

Definition 5 (Norms). *Let cond, ϕ, d be boolean combinations of propositional variables from Π and san $\in \Pi$. A conditional obligation is represented by a tuple (cond, $O(\phi)$, d, san) and a conditional prohibition is represented by a tuple (cond, $P(\phi)$, d, san). A norm set N is a set of conditional obligations and conditional prohibitions.*

Conditional norms are evaluated on runs of the physical transition system. A conditional norm $n = (cond, Y(\phi), s, san)$, where Y is O or P, is *detached* in a state satisfying its condition *cond*. Detached norms persist as long as they are not obeyed or violated, even if the triggering condition of the corresponding conditional norm does not hold any longer. A detached obligation $(cond, O(\phi), d, san)$ is *obeyed* if no state satisfying d is encountered before execution reaches a state satisfying ϕ, and *violated* if a state satisfying d is encountered before execution reaches a state satisfying ϕ. Conversely, a detached prohibition $(cond, P(\phi), d, san)$ is obeyed if no state satisfying ϕ is encountered before execution reaches a state satisfying d, and violated if a state satisfying ϕ is encountered before execution reaches a state satisfying d. If a detached norm is violated in a state s, the sanction corresponding to the norm is applied (becomes true) in s.

We say that a detached norm is annulled in a state s' immediately after a state s in which the norm is obeyed or violated, unless the same norm is detached again in s'. Note that given a state s in a transition system, we cannot say whether a norm is violated in s; to determine that, we need to know the path taken to reach s (e.g., whether any norms were detached in the past), and there may be more than one path to s. This is the reason why conditional norms are evaluated on runs of the system rather than in states.

Violation conditions of conditional norms can be expressed in temporal logic LTL $+Past$ as follows.

Definition 6 (Norm Violation). *A state $\rho[i]$ violates a conditional obligation (cond, $O(\phi)$, d, san) on run ρ in $T(M)$ iff*

$$T(M), \rho, i \models d \wedge \neg\phi \wedge ((X^{-1}(\neg\phi \wedge \neg d)\, \mathcal{S}\, (cond \wedge \neg\phi \wedge \neg d)) \vee cond)$$

$\rho[i]$ *violates a conditional prohibition (cond, $P(\phi)$, d, san) iff*

$$T(M), \rho, i \models \phi \wedge \neg d \wedge ((X^{-1}(\neg\phi \wedge \neg d)\, \mathcal{S}\, (cond \wedge \neg\phi \wedge \neg d)) \vee cond)$$

Note that whether $\rho[i]$ violates a norm is determined by the prefix of ρ ending in $\rho[i]$, and is not dependent on the future of $\rho[i]$.

3.3.2 Expressing Norms by Temporal Logic Formulas

Norm violation conditions can be expressed directly by a formula of some temporal logic. Instead of specifying e.g., conditional obligations and prohibition and then expressing their violation conditions in temporal logic, we can say that all states or all runs satisfying a temporal logic formula ϕ are prohibited. For group norms, ATL can be used to express norm violation conditions.

3.3.3 Norms as Team Plans

A history may also result from or an obligation to achieve some state or to carry out some actions by a group of agents [44]. For example, an obligation on hospital staff may require two nurses to be on duty during a particular shift. In [44], such obligations are expressed in Propositional Dynamic Logic (PDL) (see, for example, [45]).[3] In [9], similar group obligations are specified in LTL with $done(a, i)$ atoms, where $done(a, i)$ stands for 'action a has just been performed by agent i'.

3.3.4 Norms as Defeasible Rules

In [42], defeasible rules are used to specify conflicting norms, in particular, contrary-to-duty and permissive norms. In its basic form, a contrary-to-duty norm consists of a primary norm and a secondary norm, which comes into effect when the primary norm is violated. An example of a basic contrary-to-duty norm is the obligation of a customer to pay an invoice within 7 days, and if the customer does not pay the invoice within 7 days, then the

---

[3]PDL is yet another logic for describing labelled transition systems, which we did not cover in Section 2 in the interests of brevity.

customer should pay the invoice plus 5 percent interest within 15 days. In a general case, a contrary-to-duty norm consists of a sequence of norms such that when the first norm in the sequence is violated, then the second norm is in force, but if the first two norms are violated, then the third norm is in force, etc. An example of the general case of contrary-to-duty norm is the obligation of a customer to pay an invoice within 7 days, and if the customer does not pay the invoice within 7 days, then the customer should pay the invoice plus 5 percent interest within 15 days, and if the customer does not pay the invoice plus 5 percent interest within 15 days, then the customer should pay the invoice with 10 percent interest within one month. Permissive norms are exceptions to obligations and prohibitions, and an explicit permissive norm is seen as an explicit derogation of an obligation or a prohibition. For example, a general prohibition regarding the use of private protected personal data can be derogated with a permission in the sense that the permission makes an exception to the general prohibition. A contrary-to-duty or permissive norm is specified by a defeasible rule, indexed by an obligation or permission, where the consequent consists of an ordered sequence of obligations or permissions. A contrary-to-duty norm has the general form $a \Rightarrow_O b \otimes c$ and is read as "in case a holds, then b obliged, but if the obligation b is not fulfilled, then the obligation c is activated and in force". The example contrary-to-duty norm above can be represented as follows:

$$invoice \Rightarrow_O payin7days \otimes pay + 7\%in15days \otimes pay + 10\%in30days$$

A permissive norm has the general form $a \Rightarrow_P b \odot c$. Such a rule can be used to represent permissions of the type "in situation a, the subject is entitled, in the order of preference, to option b or option c". However, the reading of permissive norms is slightly different from the reading of contrary-to-duty norms since permissions cannot be violated, i.e., we cannot read the permissive norm by means of "if permission b is violated". In [42] it is argued that in the case of a permissive norm, one can proceed in the chain from b to c whenever $O\neg b$ holds. The preference operator $\odot$ establishes a preference order among permissions, and in case the opposite obligation is in force, another permission holds. In the next section, we show that the violation conditions of contrary-to-duty norms can also be expressed in temporal logic.

3.4 Expressibility of Norms in Temporal Logic

In the previous sections, we have shown that the main classes of norms described in the literature can be naturally treated as conditions on runs or histories. Such conditions can be specified in a suitable temporal logic, for example, Linear Time Temporal Logic (LTL). Depending on the goals of the specification and verification process, we can either use LTL to define the set of runs which obey the norm, or to define the set of runs which violate the norm (one is simply a negation of another).

Recently, doubts were raised in [41] regarding suitability of LTL and other temporal logics for expressing 'real life' norms. Basically, what the argument in [41] really shows is that a translation of deontic notions such as obligations and permissions into temporal logic which interprets 'obligatory' as 'always true' and 'permitted' as 'eventually true' does not work, as could be expected. However, [41] is now often cited as an argument against using temporal logic for specifying norms in general. We would like to revisit the example which is considered paradoxical when specified in LTL in [41], and show that it is possible to exactly specify the set of conditions on runs which satisfy the norms from the example using standard LTL.

The example is as follows (we compress it slightly without changing the meaning, and use the same variable names for propositions):

1. collection of personal information (A) is forbidden unless authorised by the court (C)

2. The destruction of personal information collected illegally before accessing it (B) excuses the illegal collection

3. collection of medical information (D) is forbidden unless collection of personal information is permitted

As pointed out in [41], this classifies possible situations as compliant and non-compliant as follows:

- situations satisfying C are compliant

- situations not satisfying C, where A happens but B happens as well, are weakly compliant (or correspond to a small violation; in the setting of conditional norms, this would deserve a small sanction)

- situations where C is false, where A happens and B does not, are violations

- situations not satisfying C where D happens are violations

- situations not satisfying C but also not satisfying A and D are compliant

The classification above is not very precise, since A, B, C, and D are treated as state properties which are true or false at the same time. Later in [41] a temporal relation between A and B is introduced: if C is false and A happens, then B should happen some time after that to compensate for the violation of A[4].

Hence it is very easy to classify runs into compliant or violating in LTL:

---

[4]It would have perhaps been better not to treat B as a state property, but as a property of a run, 'data not accessed until destroyed', which is expressible as $\neg Read\,\mathcal{U}\,Destroyed$, but we will stick with the formalisation in [41] to make comparison easier. Another issue is that instead of requiring B to happen 'eventually', in real life there would be some time limit on when it should happen (such as in the next state).

- Fully compliant runs:

$$\mathcal{G}(C \vee (\neg C \wedge \neg A \wedge \neg D))$$

(everywhere, either there is a court authorisation, or there is no collection of personal or medical information)

- Weakly compliant runs:

$$\mathcal{F}(\neg C \wedge A) \wedge \mathcal{G}(\neg C \wedge A \rightarrow \mathcal{F}B) \wedge \mathcal{G}(\neg C \rightarrow \neg D)$$

(there is at least one violation of prohibition on collection of personal information, but each such violation is compensated by B in the future; there are no violations of prohibition on collecting medical information)

Finally, violations are specified as follows:

- Violating runs:

$$\mathcal{F}(\neg C \wedge (D \vee (A \wedge \neg \mathcal{F}B)))$$

Note that the three formulas above define a partition of all possible runs. Clearly, there is nothing paradoxical in this specification of the set of norms.

For the sake of completeness we reproduce here the formalisation of the same set of norms in [41] and analyse where the paradoxical results come from. The set of norms is formalised in [41] as follows:

N1 $\neg C \rightarrow (\neg A \otimes B)$

N2 $C \rightarrow \mathcal{F}A$

N3 $\mathcal{G}\neg A \rightarrow \mathcal{G}\neg D$

N4 $\mathcal{F}A \rightarrow \mathcal{F}D$

N1 is intended to say that B compensates for a violation $\neg C \wedge A$. It uses a connective $\otimes$ which was introduced for expressing contrary to duty obligations. The truth definition for $\otimes$ as given in [41] is

$TS, \sigma \models \phi \otimes \psi$ iff $\forall i \geq 0, TS, \sigma_i \models \phi$ or $\exists j, k : 0 \leq j \leq k, TS, \sigma_j \models \neg \phi$ and $TS, \sigma_k \models \psi$, where TS is a transition system, and σ a run in TS. This makes $\otimes$ equivalent to

$$\mathcal{G}\phi \vee \mathcal{F}(\neg \phi \wedge \mathcal{F}\psi)$$

This condition is similar to our characterisation of weakly compliant runs, although it is stated as a property which should be true for all runs. The condition **N2** is one of the really problematic ones. It aims to say that if C holds, then A is permitted; 'permitted' is identified with 'will eventually happen'. It is quite clear that permission of A cannot be expressed as 'A will eventually happen'; the two have completely different meanings. This does not mean that LTL cannot be used for specifying norms, it just means that this particular way of specifying norms in LTL is inappropriate. **N4** is problematic in the same way: instead of saying that an occurrence of D is not a violation under the same conditions as when an occurrence of A is not a violation, it says that if A is going to happen then D is going to happen – which is again a completely different meaning. **N3** attempts to say that if A is prohibited then D is prohibited. However, instead it implies that if A happens (the antecedent $\mathcal{G}\neg A$ is false) then it does not matter whether D happens (the implication is still true). Given this formalisation, which is inappropriate in multiple ways, [41] produces an example run where **N1–N4** are true and the prohibition on collecting medical information is violated. The run consists of just two states t_1, t_2:

$$t_1 \models \neg C \wedge A \wedge D$$

$$t_2 \models B$$

which is a weakly compliant run as far as violating prohibition of A is concerned, but a non-compliant run as far as violating the prohibition of D is concerned. With our LTL specification of the set of norms it is classified as a violating run since $\mathcal{F}(\neg C \wedge D)$ holds on it. It does satisfy **N1–N4**, but clearly the problem is with **N1–N4** rather than with the intrinsic difficulty of classifying norm violating patterns in temporal logic.

3.5 Programming Norms

Another approach to specifying norms is directly in terms of programming constructs. The specification of norms can either be endogenous, i.e., form part of the programs of the (norm-compliant) agents comprising the MAS, or exogenous, i.e., form part of the program of some form of organisational framework or middleware. In these approaches, what it means for a norm to be violated is ultimately reducible to the operational semantics of the program, framework or middleware which operationalises the normative programming constructs and defines all norm-compliant executions of the normative MAS. In this section we briefly survey some of the main approaches in the literature and classify them in terms of whether they can express state, transition or history based norms.

An approach that integrates norms in a BDI-based agent programming architecture is proposed in [57]. This extends the AgentSpeak(L) architecture with a mechanism that allows agents to behave in accordance with a set of non-conflicting path-based norms. The agents can adopt obligations and prohibitions with deadlines, after which plans are selected to fulfil the obligations or existing plans are suppressed to avoid violating prohibitions.

There has also been considerable work on normative programming frameworks and middleware to support the development of normative multi-agent organisations, and such frameworks are often designed to inter-operate with existing BDI-based agent programming languages. The AMELI [35] middleware is based on the ISLANDER formal framework [34]. ISLANDER is a modelling language for specifying institutions in terms of institutional rules and norms. AMELI facilitates agent participation within the institutional environment and supports regimentation of norms relating to agents' communication actions. AMELI is thus restricted to expressing (a particular type of) transition-based norms. Other approaches, e.g., [35, 38, 63, 39], support more general action-based norms, and prescribe actions that should or should not be performed by agents. $\mathcal{S}$-$\mathcal{M}$OISE$^+$ provides support for normative MAS based on the MOISE organisational model. In $\mathcal{M}$OISE$^+$, a deontic specification states a role's permissions and obligations for missions (sets of goals). An organisational manager agent ensures that agent actions (e.g., committing to a mission) do not violate organisational constraints, including norms. However, while $\mathcal{S}$-$\mathcal{M}$OISE$^+$ provides an API which allows agents to discover their obligations, violation of obligations is not monitored by the organisational manager. JaCaMo is similar to $\mathcal{S}$-$\mathcal{M}$OISE$^+$. In JaCaMo, the organisational infrastructure of a multiagent system consists of organisational artefacts and agents that together are responsible for the management and enactment of the organisation. An organisational artefact employs a normative program which in turn implements a $\mathcal{M}$OISE$^+$ specification. Other frameworks such as ORA4MAS [49] provide support for both norm regimentation and enforcement, however monitoring and enforcement must be explicitly coded in organisational artefacts.

Other norm-based programming languages have been proposed that use high-level norms to represent what the agents should establish or should avoid, in terms of a declarative description of a system state, rather than specifying which actions actions should or should not be performed. One such language is the Organisation Oriented Programming Language (2OPL) for the implementation of normative organisations [65, 32]. In this approach, an organisation is viewed as a software entity that exogenously coordinates the interaction between agents and their shared environment. 2OPL provides programming constructs to specify 1) the initial state of an organisation, 2) the effects of agents' actions in the shared environment, and 3) the applicable norms and sanctions. In 2OPL norms can be either enforced by means of sanctions or regimented. The interpreter of 2OPL is based on a cyclic control process. At each cycle, the observable actions of the individual agents (i.e., communication and environment actions) are monitored, the effects of the actions are determined, and norms and sanction are imposed if necessary. An advantage of 2OPL approach is its complete operational semantics such that normative organisation programs can be formally analysed by means of verification techniques (see, e.g., [15, 31, 8]). A number of normative programming languages have recently been proposed that are similar in spirit to the 2OPL language. The normative language of the THOMAS multi-agent architecture [30] supports conditional norms with deadlines, sanctions and rewards. Conditions refer to actions (and optionally states). Norms are enforced rather than regulated, and sanctions may be applied by agents rather than the organization. The normative infrastructure does not restrict interactions between agents. A rule-based system implemented in Jess maintains a fact base representing the organisational state, detects norm activation and monitors violations. NPL/NOPL [47] allows the expression of norms with conditions, obligations and deadlines, and norms may be regimented or enforced. Sanctions are represented as an obligation that an agent apply the sanction to the agent that violated the norm. A translation of $\mathcal{M}OISE^+$ specifications into NOPL programs is described in [48].

4 Norm Verification

Verification of norms involves a variety of questions, answers to which all rely on the specification of norms. These questions include:

- Is a given set of norms consistent [36]? If not, compute a maximal consistent subset of this set [7].

- Given a transition system and a set of norms, are any of the norms violated [8]?

- A variant of the question above, called runtime norm monitoring, see for example [6]: given a current finite run of the system (in other words, given a finite history of the system so far), are any norms violated or about to be violated?

- Verification of the effect of applying the norms [2, 8]: given a transition system M and a set of norms N, after the norms are enforced on M, does some system objective ϕ hold in the resulting transition system? The result of applying N to M is called an implementation of a normative systems on M in [2] and is called a normative update of M with N in [8].

4.1 Norm Consistency

In this section we consider the problem of whether a set of norms are consistent. This is the focus of, for example, [36]. The authors of [36] consider two kinds of norms in electronic institutions, integrity norms which prohibit some actions after some condition occurs, and obligations, which make some actions obligatory after some condition occurs. Both are expressed in first order logic. In order to verify that an electronic institution is norm consistent, a 'dialogue' (essentially, a record of interactions between agents) must be found where there are no violations of integrity norms and there are no pending obligations. The verification problem is decidable when the domain of the ontology describing the institution is finite, so the norms can be propositionalised and the problem of checking consistency reduced to theorem proving in propositional logic.

The problem of finding a maximal consistent set of obligations arises in approaches such as BOID [23] and in the decision making mechanism of the normative programming language N2APL [7]. Essentially, the problem of finding a maximally consistent set of norms (or norms and goals) arises when a rational agent needs to decide which course of action to commit to (since it cannot commit to an inconsistent set, and at the same time may wish to obey as many norms as possible while achieving as many goals as possible). Under certain assumptions, in N2APL this problem is solvable in polynomial time (it is reduced to checking whether a certain set of plans with durations and deadlines can be scheduled in the available time).

4.2 Norm Compliance

In this section, we consider the problem: 'given a structure M and a set of norms N, are there any norm violations in M?'. If the set of norms is given semantically, we simply check whether any of the semantic conditions hold in M (are there any violating states or transitions; note that a set of violating runs even if given semantically needs to be represented in a finite way, e.g. by an automaton or by a regular expression). This problem arises as part of the problem of normative update in [8] (before sanctions could be applied, all norm violations need to be found).

If the set of norms is specified syntactically, and the set of formulas N' describes violation conditions of N, we have a model-checking problem [28, 11, 16]: for each formula ϕ in N', does M satisfy ϕ?

The model-checking problem is, given a transition system M and a formula ϕ, does $M \models \phi$ hold? The model-checking problem for different temporal logics has different complexity. The model-checking problem for CTL can be solved in time $O(|M| \times |\phi|)$; it is PTIME-complete. The model-checking problem for LTL is PSPACE-complete. It can also be solved in time $2^{O(|\phi|)} \times O(|M|)$, that is, exponential in the formula and linear in the size of the transition system, which corresponds to a more practical model-checking method than the PSPACE algorithm. The model-checking problem for CTL^* is PSPACE-complete. It can also be solved in time $2^{O(|\phi|)} \times O(|M|^2)$. The model-checking problem for ATL is PTIME-complete.

4.3 Runtime Norm Verification

According to Bauer et al. [18], runtime verification deals with those verification techniques that allow checking whether an execution of a system under scrutiny satisfies or violates a given correctness property. The problem of runtime verification is often formulated as follows: given a system to be checked and a correctness property, check whether an execution of the system satisfies or violates the correctness property. The process of runtime verification consists of various stages such as monitor synthesis, system instrumentation, and execution analysis. In the first stage, the correctness property is used to generate a monitor, which is basically a decision procedure for the property. In the second stage, relevant events of the system are fed into the monitor, and finally in the third stage, the system execution is analysed to decide whether the correctness property is satisfied or violated.

There are a variety of different formalisms that are proposed in the literature to specify and develop monitors that encode the correctness properties. These proposals varies from runtime verification specific formalisms to general purpose formalisms. Some RV-domain specific formalisms, as listed in [18], are language oriented formalisms such as extended regular expressions [62], tracematches by the ApectJ team [10], query-oriented languages such as PQL [56], and rule-based approaches [17]. More generic and general purpose formalisms to specify and develop monitors are various fragments of linear temporal logic [40, 46, 64], various types of automata such as security automata [59] for encoding safety properties and edit automata [55] for encoding non-safety properties, and aspect-oriented programming such as AspectJ that can be used to develop monitors.

Assuming that correctness properties are closely related to norms in the sense that both are properties that system executions can satisfy or violate, techniques from runtime verification can be used to check norm violations at runtime. For runtime norm verification, the monitor synthesis stage is most relevant as it encodes a norm to a monitor that is subsequently used to decide violation/satisfaction of the norm at runtime. In the following, we present some of the general formalisms from runtime verification literature that can be used to encode norms for runtime norm monitoring purposes.

4.3.1   Runtime Verification for LTL-based norms

As mentioned above, various fragments of LTL are used to specify norms. In standard LTL, a formula specifies a property of infinite runs. However, following [18], the goal of runtime verification is to check properties given finite prefixes of infinite runs. Given that norms are specified as LTL formula, runtime norm verification should check whether finite prefixes of infinite runs are compliant or violate norms. For runtime verification of LTL properties, [18] proposes a three valued semantics. Adopting this semantics for norms, a finite run can be norm compliant, norm violating, or inconclusive in the sense that the norm cannot be said to be satisfied or violated. In general, given a finite run r, a norm n is violated if there is no continuation of r that satisfies n, satisfied if all possible continuations of r satisfy n, and inconclusive otherwise. Formally, let Π be a set of atomic propositions, $\Sigma = 2^{\Pi}$ be an finite alphabet, Σ^{ω} be the set of all infinite words (runs), and Σ^* be the set of all finite words (runs), $r\sigma \in \Sigma^{\omega}$ be an infinite run starting with finite prefix $r \in \Sigma^*$ followed by infinite run

$\sigma \in \Sigma^\omega$, and $\models_{LTL}$ be the standard LTL satisfaction relation.

r satisfies n	if $\forall \sigma \in \Sigma^\omega : r\sigma \models_{LTL} n$
r violates n	if $\forall \sigma \in \Sigma^\omega : r\sigma \not\models_{LTL} n$
r is inconclusive wrt n	otherwise

Given that arbitrary LTL formula can be evaluated on finite runs, [18] describe the construction of a (deterministic) finite state machine that can read a finite run and determine whether it satisfies, violates, or is inconclusive with respect to a LTL property.

As shown in [18], the size of the resulting monitor is double exponential in the size of $|\phi|$ and the cause of this is related to the construction of the Büchi automata and the construction of the product automaton.

One interesting characteristic of their construction is that the satisfaction and violations of properties can be decided as early as possible. This feature is particularly important for adopting this approach for runtime monitoring of norms. It should be noted that the adoption of this approach for runtime verification of norms is limited to the detection of norm violations and cannot deal with norm enforcement to regiment or sanction violations (see e.g., [8, 6]).

4.3.2 Runtime Enforcement for Safety-Progress Properties

Norms can be enforced on a system by means of regimentation or sanctions. In the first case, the violation of norms are prevented by either ignoring/undoing the violating actions or by halting/blocking the execution of the system. In case of sanctioning, norm violation are allowed but compensated by intervening in the system run. In the context of norm enforcement, the specification of norms is not only for the monitoring purposes, but also for the intervention. The specification of a norm should therefore include a regiment/sanction modality, and in the case of sanctioning, also the sanction that should be imposed upon the norm violation. In the field of runtime verification, mechanisms are devised to enforce properties at runtime [59, 54, 55]. Examples of these mechanisms are truncation, suppression, insertion, and edit automata. These automata are known under the general term security automata that are designed to enforce security properties. The properties that security automata can enforce are specified with respect to the general Safety-Progress classification of properties [27].

A truncation automaton is defined with respect to a system and can be seen as a sequence recogniser and is designed to halt the system run when the system attempts to invoke a forbidden operation. Such an automaton is defined as a finite or countably infinite state machine and with respect to a set of actions of the system under scrutiny. The transition function of a truncation automaton is a partial function and indicates whether to accept the current operation of the system under scrutiny and move to a new state or to halt the target program. Formally, a truncation automaton is tuple (Q, q_0, δ) defined with respect to a system with action set $\mathcal{A}$, where Q is the set of possible states of the automaton, $q_0 \in Q$ is the initial state of the automaton, and $\delta : \mathcal{A} \times Q \to Q$ is a partial transition function that determines which system actions to be accepted. The operational semantics of the truncation automaton is defined by the following transition rules:

$$
\begin{aligned}
(a; \sigma , q) &\xrightarrow{a} (\sigma , q') &&\text{if } \delta(a, q) = q' &&\text{(STEP)} \\
(\sigma , q) &\xrightarrow{\cdot} (\cdot , q) &&\text{otherwise} &&\text{(STOP)}
\end{aligned}
$$

The transition rule STEP accepts the current system action a allowing the system run to proceed and the transition rule STOP halts the system run. In [37], it is shown that the class of properties that truncation automaton can enforce is the class of safety properties of the form $\mathcal{G}\phi$, where ϕ is a past formula.

Edit automata are more powerful and extend truncation automata. An edit automaton is defined as a finite or countably infinite state machine and with respect to a set of actions of the system under scrutiny. The transition function of an edit automaton includes the partial transition function of the truncation automaton, but add two new partial transition functions with disjoint domains. One partial transition function indicates whether or not an operation of the system under scrutiny should be suppressed or accepted, while the second one specifies the insertion of a finite sequence of operations to be inserted in the run of the system under scrutiny. An edit automaton can thus allow, suppress, or halt the execution of the system under scrutiny or even insert finite sequences of operations in the execution of the system. Formally, an edit automaton is tuple $(Q, q_0, \delta, \gamma, \omega)$ defined with respect to a system with action set $\mathcal{A}$, where $Q, q_0 \in Q$ and δ are defined as with the truncation automata, $\gamma : \mathcal{A} \times Q \to \vec{A} \times Q$ is a partial function that specifies the insertion of a finite sequence of actions into the system run, and $\omega : \mathcal{A} \times Q \to \{-, +\}$ is a partial function that specifies whether system actions should be accepted or suppressed. In edit automata it is assumed that δ and ω have the same domain, while δ and γ have disjoint domains. The operational semantics of the edit automaton is defined by the following transition rules:

$$
\begin{aligned}
(a; \sigma , q) &\xrightarrow{a} (\sigma , q') &&\text{if } \delta(a, q) = q' \text{ and } \omega(a, q) = + &&\text{(STEPA)} \\
(a; \sigma , q) &\xrightarrow{\cdot} (\sigma , q') &&\text{if } \delta(a, q) = q' \text{ and } \omega(a, q) = - &&\text{(STEPS)} \\
(a; \sigma , q) &\xrightarrow{\tau} (a; \sigma , q') &&\text{if } \gamma(a, q) = (\tau, q') \text{ and } \omega(a, q) = - &&\text{(INS)} \\
(\sigma , q) &\xrightarrow{\cdot} (\cdot , q) &&\text{otherwise} &&\text{(STOP)}
\end{aligned}
$$

The transition rule STEPA accepts the current system action a allowing the system run to proceed, the transition rule STEPS suppresses the current system action a but allows the system run to proceed, the transition rule INS adds a finite sequence of actions τ and allows the system run to proceed, and finally the STEP transition rule halts the system run.

In [37], it is shown that the class of properties that edit automaton can enforce is the class of response properties of the form $\mathcal{G}\mathcal{F}\phi$, where ϕ is a past formula.

4.3.3 Runtime Norms Verification with Aspect Oriented Programming

Aspect-oriented programming is an extension to object-oriented programming and allows software developers to create software systems that can grow to meet changing requirements. This is done by supporting dynamic modifications of software systems without changing their static object-oriented model. The dynamics modifications can be realised by including some new code to satisfy changing requirements in a separate single location rather than incorporating it at various locations in the existing software. Aspect oriented programming also allows to extend software systems to satisfy new requirements even if the code of the software system is not available. The key concepts of aspect oriented programming are point-cut and advice. An object-oriented program exposes joint points which can be selected by pointcuts. Such join points are points of execution in the software application where an intervention has to be realised in order to meet the new requirements. For example, a point-cut may refer to the point just before or after the execution thread enters or exits a method of some object. An advice is the new code that should be added to the existing object-oriented model to ensure the new requirements. This additional code implements the intervention in the execution of the existing object-oriented software application. A simple example of an advice is the logging code that a developer wants to apply just before or after the execution thread enter or exits a method of some object. A point-cut together with a corresponding advice is then called an aspect. The most mature aspect oriented programming language is AspectJ, which is an extension of Java.

Following the parallel with runtime verification, one may use aspect oriented programming to specify norms and enforce them during the execution of a software system. In the object oriented programming paradigm, state-based norms can be specified in terms of some state variables, action-based norms can be specified in terms of method calls, and behavioural norms can be implemented by creating some additional data structures. The enforcement of norms can be realised by means of point-cuts referring to the execution points where the value of some state variables change, when some method is called, or a combination thereof with some additional data structures. In particular, a norm can be specified by means of some point-cuts with corresponding queries on values of state-variables or arguments of method calls. The enforcement of a norm by means of regimentation or sanctioning can be modelled as an aspect that combines some point-cut and an advice. In particular, the violation of norms and possible sanctions can be implemented in the advice of an aspect.

For example, consider a behavioural/temporal norm that is specified in terms of a condition, an obligation/prohibition and a deadline. Such a norm specifies that some state should be reached or prevented (or some actions should be performed or avoided), as soon as the specified condition is met and before the deadline is reached. The violation of a norm applies an intervention procedure that in case of norm regimentation halts the software execution or in case of norm sanctioning imposes a sanction. A conditional norm can be implemented by means of a number of related pointcuts and advice pairs, in particular, one pair for the condition of the norm, one for the content of the norm (obligation or prohibition), and one for the deadline of the norm.

When the condition pointcut of the norm is reached, then the condition advice checks whether the norm should be instantiated and detached. If so, then a detached norm (obligation or prohibition) is created and stored in a specially designed data structure called detached norm list. Suppose the norm is an obligation. If the obligation pointcut is reached, then the obligation advice will check whether the stored detached obligation in the detached norm list is fulfilled and can be removed from the list. Moreover, if the pointcut of the deadline is reached, then the advice of the deadline pointcut will not proceed the call if the obligation is regimented, and otherwise executes the sanction that corresponds to the obligation. Suppose the norm is a prohibition. If the pointcut of the prohibition is reached, then its advice checks whether a detached prohibition exists. If so, then the prohibition is considered as violated. In case the prohibition is regimented, the advice will not proceed the pointcut's method call. Otherwise, if the prohibition is sanctioned, then the sanction will be executed. If the deadline pointcut is reached and the norm is a prohibition, then the detached prohibition is considered as fulfilled and removed from the detached norm list.

4.4 Normative Update

In this section, we consider the following *normative update problem*: 'given a structure and a set of norms, does the structure where this set of norms is enforced, satisfy a certain system objective?'. Formally, it can be characterised as follows. Given a model M of a computational system (e.g., a transition system) that does not satisfy an overall desirable system property ϕ (e.g., a LTL formula), decide whether the enforcement of a set of norms N (e.g., conditional norm with deadline) on M, denoted as $M \upharpoonright N$, satisfies the desirable system level property ϕ. N can be either a regimentation norm or a sanctioning norm with corresponding sanction that are imposed on violations. Thus, given $M \not\models \phi$, decide whether $M \upharpoonright N \models \phi$.

This problem has been studied for state-based, action-based, and behaviour-based norms.

In [2], norms are specified semantically as a set of prohibited transitions (edges). A set of norms implemented on a transition system M results in a new transition system, $M \upharpoonright N$, which is M with all prohibited edges removed. In [2], the 'reasonableness assumption' is made, which is that $M \upharpoonright N$ is always non-empty: no set of norms will disable all possible transitions in the system. The most basic question to ask is whether $M \upharpoonright N$ satisfies some design objective ϕ. The authors state that the same approach would work with state-based norms, where a norm correspond to a set of prohibited states. In the latter case, $M \upharpoonright N$ is M with all the states prohibited by N removed.

In [8], the notion of applying a set of norms to a transition system is studied for conditional norms with sanctions and deadlines. [8] state that their approach subsumes that of [2]. They distinguish two kinds of norms: regimenting and sanctioning norms. *Regimenting norms* can be used to ensure that certain state or behaviours never occur. If a norm labels a state with the distinguished sanction atom $san_\perp$, then the run containing this state is removed from the set of runs of the system by the normative update. *Sanctioning norms*, on the other hand, can be used to penalise rather than eliminate certain execution paths. An undesirable state (from the point of view of the system designer) may or may not be achievable by an agent (or agents) depending on the resources the agent is able or willing to commit to achieving it.

The *normative update* of a physical transition system is defined by applying sanctioning and regimenting norms to the computation tree of the system. The application of sanctioning norms changes the valuation of the violating states (sanction atoms are added), while the application of regimenting norms removes branches of the tree where one of the states violates the regimenting norm.

Definition 7 (Normative Update). *Let $M = (S, R, V, s_I)$ be a finite transition system, $T(M)$ be the computational tree of M, and N a finite set of conditional obligations and prohibitions. The normative update of $T(M)$ with N, denoted as $T^N(M)$, is obtained from $T(M)$ as follows:*

- *for every state s of $T(M)$, if s violates a sanctioning norm $n \in N$, then the sanction atom of n is added to the valuation of s*

- *all branches which contain a state violating a regimenting norm $n \in N$ are removed from $T(N)$.*

Observe that each node s' of $T^N(M)$ contains sanction atoms of all norms violated in s'. Observe also that runs which contain a state satisfying the distinguished sanction atom $san_\perp$ are removed from $T(M)$. As in [2], [8] also assume that $T^N(M)$ is non-empty, i.e., that regimentation does not remove all possible paths from the system.

To formulate system objectives, [8] introduce two logics, variants of CTL and ATL, where path quantifiers are annotated with multisets of sanctions incurred on a path. This allows them to express properties like 'when norms are enforced, the agent(s) are unable to bring about a state satisfying (a bad property) ϕ without incurring at least sanctions Z', where Z is a multiset of sanctions. For both logics, the problem of checking whether the normative update satisfies a property is PSPACE-complete.

In [53], two variants of a dynamic modal logic are proposed to characterise and reason about norm dynamics. The first variant of the logic is devised for updates with state-based norm, while the second variant is devised for updates with action-based norms. The proposed logics come with corresponding sound and complete proof systems. The logics are devised to represent norm updates and to reason about the effect of such updates on a system specification. The logics provide update operators for adding norms to a system specification and to reason about the effects of such updates on the system specification. A target system is modelled as a labelled transition system that determines the effect of actions on the system states. Motivated by the Anderson's reduction [13], violation atoms are used to label the norm violating states.

The first type of norms are state-based and have the form $(\phi, +v, Act_R)$ or $(\phi, -v, Act_R)$, where ϕ is the norm condition, $+v$ and $-v$ are the norm effect, and Act_R is a set of repair actions. Adding such a norm to a system updates the system in such a way that for every ϕ state the violation atom v holds until a repair action from Act_R occurs. The idea of the repair action is that its occurrence repairs the system violation. Adding a norm of the form $(\phi, -v, Act_R)$ has a similar effect except that the proposition v stops to hold until the norm effect is repaired. An example of such a norm is $(station, +v, \{buy\_sub\})$, which represents the norm that being in a (train) station causes violation v unless this effect is repaired by buying a subscription. Updating a system with a norm duplicates its states to create two types of states: states in which norm effects are active and states in which norm effects are repaired. The transition relation is then modified in such a way that any violating action ends up in a state where the norm effect is active, and that any repair action ends up in states where the norm effect is repaired. The second type of norms are action-based of the form $Act_T, \phi, +v, Act_r$ and $Act_T, \phi, -v, Act_r$, where the new Act_T component is the set of actions that trigger a norm in the sense that after the triggering actions for every ϕ state the violation atom v holds until a repair action from Act_R occurs. Similar reading is used for norms of the form $Act_T, \phi, -v, Act_r$. An example of this second norm type is $(unchecked, station, +v, \{leave\})$, which represents the norm that unchecked entrance of a (train) station causes violation atom v unless this effect is repaired by buying a subscription.

5 Summary

Violation conditions of regulative norms may correspond to conditions on states, actions, or arbitrary temporal patterns. They may be specified semantically or expressed syntactically in a suitable temporal logic, or in a programming language. Verification problems for norms or rather for normative systems involve verifying consistency of norms, verifying whether violation conditions hold, and finally verifying whether a system where norms are enforced satisfies some system objective. We summarise the material covered in this chapter in a table below.

Specification	Verification Problems
State-based	Consistency; Compliance; Update
Action-based	Consistency; Compliance; Run-time monitoring; Run-time enforcement; Update
Behaviour-based	Consistency; Compliance; Run-time monitoring; Run-time Enforcement; Update

Table 1: Summary of specification and verification of norms

References

[1] Thomas Ågotnes, Wiebe van der Hoek, and Michael Wooldridge. Normative system games. In *Proceedings of the 6th International Joint Conference on Autonomous Agents and Multiagent Systems (AAMAS '07)*, pages 1–8, New York, NY, USA, 2007. ACM.

[2] Thomas Ågotnes, Wiebe van der Hoek, and Michael Wooldridge. Robust normative systems and a logic of norm compliance. *Logic Journal of the IGPL*, 18(1):4–30, 2010.

[3] Carlos E. Alchourrón and Eugenio Bulygin. The expressive conception of norms. In Risto Hilpinen, editor, *New Studies in Deontic Logic*, pages 95–124. OUP Oxford, 1981.

[4] H. Aldewereld, V. Dignum, and W. Vasconcelos. We ought to; they do; blame the management! – a conceptualisation of group norms. In *Proc. 15th Int. Workshop on Coordination, Organisations, Institutions and Norms (COIN 2013)*, 2013.

[5] Huib Aldewereld, Sergio Álvarez-Napagao, Frank Dignum, and Javier Vázquez-Salceda. Engineering social reality with inheritance relations. In Huib Aldewereld, Virginia Dignum, and Gauthier Picard, editors, *Engineering Societies in the Agents World X, 10th International Workshop, Proceedings ESAW 2009*, volume 5881 of *Lecture Notes in Computer Science*, pages 116–131. Springer, 2009.

[6] Natasha Alechina, Nils Bulling, Mehdi Dastani, and Brian Logan. Practical run-time norm enforcement with bounded lookahead. In *Proceedings of the 2015 International*

Conference on Autonomous Agents and Multiagent Systems, AAMAS 2015, Istanbul, Turkey, May 4-8, 2015, pages 443–451, 2015.

[7] Natasha Alechina, Mehdi Dastani, and Brian Logan. Programming norm-aware agents. In *Proceedings of the 11th International Conference on Autonomous Agents and Multiagent Systems (AAMAS 2012)*, pages 1057–1064. IFAAMAS, 2012.

[8] Natasha Alechina, Mehdi Dastani, and Brian Logan. Reasoning about normative update. In *Proceedings of the Twenty Third International Joint Conference on Artificial Intelligence (IJCAI 2013)*, pages 20–26. AAAI Press, 2013.

[9] Natasha Alechina, Wiebe van der Hoek, and Brian Logan. Fair allocation of group tasks according to social norms. In Nils Bulling, Leendert van der Torre, Serena Villata, Wojtek Jamroga, and Wamberto Vasconcelos, editors, *Computational Logic in Multi-Agent Systems, 15th International Workshop, CLIMA XV, Prague, Czech Republic, August 18-19, 2014*, volume 8624 of *Lecture Notes in Computer Science*, pages 19–34, Prague, Czech Republic, August 2014. Springer.

[10] Chris Allan, Pavel Avgustinov, Aske Simon Christensen, Laurie Hendren, Sascha Kuzins, Ondřej Lhoták, Oege de Moor, Damien Sereni, Ganesh Sittampalam, and Julian Tibble. Adding trace matching with free variables to aspectj. In *Proceedings of the 20th Annual ACM SIGPLAN Conference on Object-oriented Programming, Systems, Languages, and Applications*, OOPSLA '05, pages 345–364, New York, NY, USA, 2005. ACM.

[11] Rajeev Alur, Thomas Henzinger, and Orna Kupferman. Alternating-time temporal logic. *Journal of the ACM*, 49(5):672–713, 2002.

[12] A.R. Anderson. The formal analysis of normative concepts. *American Sociological Review*, 22:9–17, 1957.

[13] A.R. Anderson. The logic of norms. *Logique et Analyse*, 2:84–91, 1958.

[14] A.R. Anderson. A reduction of deontic logic to alethic modal logic. *Mind*, 22:100–103, 1958.

[15] L. Astefanoaei, M. Dastani, J.J. Meyer, and F. de Boer. On the semantics and verification of normative multi-agent systems. *International Journal of Universal Computer Science*, 15(13):2629–2652, 2009.

[16] Christel Baier and Joost-Pieter Katoen. *Principles of Model Checking*. The MIT Press, 2007.

[17] Howard Barringer, David Rydeheard, and Klaus Havelund. Rule systems for run-time monitoring: From Eagle to RuleR. In Oleg Sokolsky and Serdar Taşıran, editors, *Runtime Verification: 7th International Workshop, RV 2007, Vancover, Canada, March 13, 2007, Revised Selected Papers*, pages 111–125, Berlin, Heidelberg, 2007. Springer Berlin Heidelberg.

[18] Andreas Bauer, Martin Leucker, and Christian Schallhart. Runtime verification for LTL and TLTL. *ACM Transactions on Software Engineering and Methodology*, 20(4):14:1–14:68, 2011.

[19] Cristina Bicchieri. *The Grammar of Society: The Nature and Dynamics of Social Norms*. Cambridge University Press, March 2006.

[20] Cristina Bicchieri and Hugo Mercier. Norms and beliefs: How change occurs. In Maria Xenitidou and Bruce Edmonds, editors, *The Complexity of Social Norms*, pages 37–54. Springer International Publishing, 2014.

[21] Guido Boella and Leendert van der Torre. Regulative and constitutive norms in normative multiagent systems. In *Proceedings of the Ninth International Conference on Principles of Knowledge Representation and Reasoning (KR'04)*, pages 255–266, 2004.

[22] J. Broersen, M. Dastani, J. Hulstijn, and L. van der Torre. Goal generation in the BOID architecture. *Cognitive Science Quarterly*, 2(3-4):428–447, 2002.

[23] Jan Broersen, Mehdi Dastani, Joris Hulstijn, Zisheng Huang, and Leendert van der Torre. The boid architecture: Conflicts between beliefs, obligations, intentions and desires. In *Proceedings of the Fifth International Conference on Autonomous Agents*, AGENTS '01, pages 9–16, New York, NY, USA, 2001. ACM.

[24] Nils Bulling and Mehdi Dastani. Verification and implementation of normative behaviours in multi-agent systems. In *Proc. of the 22nd Int. Joint Conf. on Artificial Intelligence (IJCAI)*, pages 103–108, Barcelona, Spain, July 2011.

[25] Nils Bulling, Mehdi Dastani, and Max Knobbout. Monitoring norm violations in multi-agent systems. In *Twelfth International conference on Autonomous Agents and Multi-Agent Systems (AAMAS'13)*, pages 491–498, 2013.

[26] C. Castelfranchi. Formalizing the informal?: Dynamic social order, bottom-up social control, and spontaneous normative relations. *JAL*, 1(1-2):47–92, 2004.

[27] Edward Chang, Zohar Manna, and Amir Pnueli. The safety-progress classification. In Friedrich L. Bauer, Brauer Wilfried, and Helmut Schwichtenberg, editors, *Logic and Algebra of Specification*, volume 94 of *NATO ASI Series*, pages 143–202. Springer, 1993.

[28] E. M. Clarke, E. A. Emerson, and A. P. Sistla. Automatic verification of finite-state concurrent systems using temporal logic specifications. *ACM Transactions on Programming Languages and Systems*, 8(2):244–263, 1986.

[29] O. Cliffe, M. de Vos, and J. Padget. Specifying and reasoning about multiple institutions. In *Coordination, Organizations, Institutions, and Norms in Agent Systems II*, volume 4386, pages 67–85. Springer LNCS, 2007.

[30] Natalia Criado, Vicente JuliÃ₫n, Vicente Botti, and Estefania Argente. A norm-based organization management system. In Julian Padget, Alexander Artikis, Wamberto Vasconcelos, Kostas Stathis, VivianeTorres Silva, Eric Matson, and Axel Polleres, editors, *Coordination, Organizations, Institutions and Norms in Agent Systems V*, volume 6069 of *Lecture Notes in Computer Science*, pages 19–35. Springer Berlin Heidelberg, 2010.

[31] Mehdi Dastani, Davide Grossi, and John-Jules Meyer. A logic for normative multi-agent programs. *Journal of Logic and Computation, special issue on Normative Multiagent Systems*, 23(2):335–354, 2013.

[32] Mehdi Dastani, Nick AM Tinnemeier, and John-Jules Ch Meyer. A programming language for normative multi-agent systems. *Multi-Agent Systems: Semantics and Dynamics of Organizational Models*, pages 397–417, 2009.

[33] J. Elster. Social norms and the explanation of behavior. *The Oxford handbook of analytical sociology*, page 195âĂŞ217, 2009.

[34] M. Esteva, D. de la Cruz, and C. Sierra. ISLANDER: an electronic institutions editor. In *Proceedings of the First International Joint Conference on Autonomous Agents and MultiAgent Systems (AAMAS 2002)*, pages 1045–1052, Bologna, Italy, 2002.

[35] M. Esteva, J.A. Rodríguez-Aguilar, B. Rosell, and J.L. Arcos. AMELI: An agent-based middleware for electronic institutions. In *Proceedings of AAMAS 2004*, pages 236–243, New York, US, July 2004.

[36] Marc Esteva, Juan Rodriguez-Aguilar, Carles Sierra, and Wamberto Vasconcelos. Verifying norm consistency in electronic instixions. In Virginia Dignum, Daniel Corkill, Catholijn Jonker, and Frank Dignum, editors, *Proceedings of the AAAI-04 Workshop*

on Agent Organizations: Theory And Practice, volume Technical Report WS-04-02, San Jose, July 2004. AAAI, AAAI Press.

[37] Yliès Falcone, Laurent Mounier, Jean-Claude Fernandez, and Jean-Luc Richier. Runtime enforcement monitors: composition, synthesis, and enforcement abilities. *Formal Methods in System Design*, 38(3):223–262, 2011.

[38] A. Garcia-Camino, P. Noriega, and J. A. Rodriguez-Aguilar. Implementing norms in electronic institutions. In *Proceedings of the Fourth International Joint Conference on Autonomous Agents and MultiAgent Systems (AAMAS'05)*, pages 667–673, New York, NY, USA, 2005.

[39] A. García-Camino, J. Rodríguez-Aguilar, C. Sierra, and W. Vasconcelos. Constraint rule-based programming of norms for electronic institutions. *Autonomous Agents and Multi-Agent Systems*, 18:186–217, 2009.

[40] Dimitra Giannakopoulou and Klaus Havelund. Automata-Based Verification of Temporal Properties on Running Programs. In *Proceedings of the 16th IEEE International Conference on Automated Software Engineering*, pages 412–416. IEEE Computer Society, 2001.

[41] Guido Governatori. Thou shalt is not you will. In Ted Sichelman and Katie Atkinson, editors, *Proceedings of the 15th International Conference on Artificial Intelligence and Law, ICAIL 2015*, pages 63–68. ACM, 2015.

[42] Guido Governatori, Francesco Olivieri, Antonino Rotolo, and Simone Scannapieco. Computing strong and weak permissions in defeasible logic. *Journal of Philosophical Logic*, 42(6):799–829, 2013.

[43] D. Grossi. *Designing Invisible Handcuffs*. PhD thesis, Utrecht University, SIKS, 2007.

[44] D. Grossi, F. Dignum, L. M. M. Royakkers, and J-J. Ch. Meyer. Collective obligations and agents: Who gets the blame? In *Deontic Logic in Computer Science*, volume 3065, pages 129–145. Springer LNCS, 2004.

[45] David Harel. Dynamic logic. In D. Gabbay and F. Guenthner, editors, *Handbook of Philosophical Logic, Volume II: Extensions of Classical Logic*, volume 165 of *Synthese Library*, chapter II.10, pages 497–604. D. Reidel Publishing Co., Dordrecht, 1984.

[46] Klaus Havelund and Grigore Rosu. Efficient monitoring of safety properties. *International Journal on Software Tools for Technology Transfer*, 6(2):158–173, August 2004.

[47] Jomi Hübner, Olivier Boissier, and Rafael Bordini. A normative programming language for multi-agent organisations. *Annals of Mathematics and Artificial Intelligence*, 62:27–53, 2011.

[48] Jomi Fred Hübner, Olivier Boissier, and Rafael H. Bordini. From organisation specification to normative programming in multi-agent organisations. In Jürgen Dix, João Leite, Guido Governatori, and Wojtek Jamroga, editors, *Computational Logic in Multi-Agent Systems, 11th International Workshop, CLIMA XI, Lisbon, Portugal, August 16-17, 2010. Proceedings*, volume 6245 of *Lecture Notes in Computer Science*, pages 117–134. Springer, 2010.

[49] JomiF. Hübner, Olivier Boissier, Rosine Kitio, and Alessandro Ricci. Instrumenting multi-agent organisations with organisational artifacts and agents. *Autonomous Agents and Multi-Agent Systems*, 20:369–400, 2010.

[50] A. J. I. Jones and M. Sergot. On the characterization of law and computer systems. In J.-J. Ch. Meyer and R.J. Wieringa, editors, *Deontic Logic in Computer Science: Normative System Specification*, pages 275–307. John Wiley & Sons, 1993.

[51] S. Kanger. New foundations for ethical theory. In R. Hilpinen, editor, *Deontic Logic: Introductory and Systematic Readings*, pages 36–58. Reidel Publishing Company, 1971.

[52] Max Knobbout and Mehdi Dastani. Reasoning under compliance assumptions in normative multiagent systems. In Wiebe van der Hoek, Lin Padgham, Vincent Conitzer, and Michael Winikoff, editors, *Proceedings of the 11th International Conference on Autonomous Agents and Multiagent Systems (AAMAS 2012)*, pages 331–340. IFAAMAS, 2012.

[53] Max Knobbout, Mehdi Dastani, and John-Jules Meyer. A dynamic logic of norm change. In *22nd European Conference on Artificial Intelligence, The Hague, The Netherlands, August 29 - September 2, 2016*. To be published, 2016.

[54] Jay Ligatti, Lujo Bauer, and David Walker. Edit automata: Enforcement mechanisms for run-time security policies. *International Journal of Information Security*, 4(1-2):2–16, 2005.

[55] Jay Ligatti, Lujo Bauer, and David Walker. Run-time enforcement of nonsafety policies. *ACM Trans. Inf. Syst. Secur.*, 12(3):19:1–19:41, January 2009.

[56] Michael Martin, Benjamin Livshits, and Monica S. Lam. Finding application errors and security flaws using PQL: A program query language. In *Proceedings of the*

20th Annual ACM SIGPLAN Conference on Object-oriented Programming, Systems, Languages, and Applications, OOPSLA '05, pages 365–383, New York, NY, USA, 2005. ACM.

[57] Felipe Rech Meneguzzi and Michael Luck. Norm-based behaviour modification in BDI agents. In Carles Sierra, Cristiano Castelfranchi, Keith S. Decker, and Jaime Simão Sichman, editors, *Proceedings of the 8th International Joint Conference on Autonomous Agents and Multiagent Systems (AAMAS 2009)*, pages 177–184. IFAAMAS, 2009.

[58] Amir Pnueli. The temporal logic of programs. In *Proceedings of the 18th Annual Symposium on Foundations of Computer Science (FOCS)*, pages 46–57, 1977.

[59] Fred B. Schneider. Enforceable security policies. *ACM Trans. Inf. Syst. Secur.*, 3(1):30–50, February 2000.

[60] Ph. Schnoebelen. The complexity of temporal logic model checking. In Philippe Balbiani, Nobu-Yuki Suzuki, Frank Wolter, and Michael Zakharyaschev, editors, *Advances in Modal Logic 4*, pages 393–436. King's College Publications, 2003.

[61] J. Searle. *The Construction of Social Reality*. Free, 1995.

[62] Koushik Sen and Grigore RoÅ§u. Generating optimal monitors for extended regular expressions. *Electronic Notes in Theoretical Computer Science*, 89(2):226 – 245, 2003. {RV} '2003, Run-time Verification (Satellite Workshop of {CAV} '03).

[63] V. Torres Silva. From the specification to the implementation of norms: an automatic approach to generate rules from norms to govern the behavior of agents. *International Journal of Autonomous Agents and Multiagent Systems (JAAMAS)*, 17(1):113–155, 2008.

[64] Volker Stolz and Eric Bodden. Temporal assertions using aspectj. *Electronic Notes in Theoretical Computer Science*, 144(4):109 – 124, 2006. Proceedings of the Fifth Workshop on Runtime Verification (RV 2005)Proceedings of the Fifth Workshop on Runtime Verification.

[65] Nick Tinnemeier, Mehdi Dastani, John-Jules Meyer, and Leon van der Torre. Programming normative artifacts with declarative obligations and prohibitions. In *Proceedings of the IEEE/WIC/ACM International Conference on Intelligent Agent Technology (IAT'09)*, pages 69–78, 2009.

Received 24 June 2016

MODELING NORM DYNAMICS IN MULTI-AGENT SYSTEMS

CHRISTOPHER K. FRANTZ
Norwegian University of Science and Technology (NTNU)
Department of Computer Science (IDI)
2815 Gjøvik, Norway
cf@christopherfrantz.org

GABRIELLA PIGOZZI
Université Paris-Dauphine, PSL Research University
CNRS, LAMSADE
75016 Paris, France
gabriella.pigozzi@dauphine.fr

1 Introduction and Motivation

Since multi-agent systems are inspired by human societies, they do not only borrow their coordination mechanisms such as conventions and norms, but also need to consider the processes that describe *how norms come about*, *how they propagate in the society*, and *how they change over time*.

In the NorMAS community, this is best reflected in various norm life cycle conceptions that look at normative processes from a holistic perspective. While the earliest life cycle model emerged in the research field of international relations, the first life cycle model in the AI community has been proposed at the 2009 NorMAS Dagstuhl workshop by Savarimuthu and Cranefield [2009] and is based on a comprehensive survey of then existing contributions to the research field. Subsequently, two further models have been proposed that offer more refined accounts of the fundamental underlying processes.

In this article, we review all existing norm life cycle models (Section 2), including the introduction of the individual life cycle models and their contextualization with specific contributions that exemplify life cycle processes. In addition, we provide a *comprehensive*

We would like to thank Bastin Tony Roy Savarimuthu for bringing us together and providing us with the opportunity to collaborate on this article.

contemporary overview of individual contributions to the area of NorMAS and a *systematic comparison of the discussed life cycle models* (Section 2.6). Based on this analysis, we propose a refined *general norm life cycle model* that resolves terminological ambiguities and ontological inconsistencies of the existing models while reflecting the contemporary view on norm formation and emergence.

This comprehensive review of life cycle models represents the birds-eye perspective on dynamics in normative multi-agent systems, which is complemented by research areas that operate at the intersection of normative processes captured by life cycle models. In addition to this holistic perspective, we thus discuss two active research fields that deal with norm dynamics: norm change and norm synthesis.

In human societies, norms change over time: new norms can be created to face changes in the society, old norms can be retracted either because they became obsolete or because superseded by others, and also norms can be modified. Thus, multi-agent systems too need mechanisms to model and reason about norm change. The field of *norm change* (Section 3) puts a specific focus on the definition of mechanisms that describe and regulate the change of norms over time. Essential aspects include the *translation of legal to logical specifications*, the *definition of a normative approach to norm change*, and the *tuning of computational mechanisms for norm change*. This research area is rather recent and to date there is still no consensus on a common account for norm change. This section retraces the historical development and debates within this field and provides an outlook on future directions.

The second subfield, *norm synthesis* (Section 4), has a longer history that has its roots in the systems engineering domain and is concerned with the use of norms and social laws as scalable coordination mechanisms in open systems. The associated challenges are twofold and have led to the development of distinct branches, with one concentrating on the analysis of factors that mitigate the *emergence of norms or conventions*, and the second one focusing on the *identification and classification of norms* in existing normative environments. This section identifies a taxonomy of norm synthesis approaches based on a comprehensive literature overview of the field, and illustrates contemporary developments using selected contributions.

We conclude this article by contextualizing the discussed subfields with the proposed general norm life cycle model, reflecting on the progression of research on norm dynamics, and finally, by providing an outlook on contemporary and future challenges of modeling of norm dynamics.

2 Norm Life Cycle Models

In the following sections, we introduce four norm life cycle models discussed in the literature to date. The models are organized chronologically, and, with exception of the last model by Mahmoud *et al.* (Section 2.4), are of increasing complexity. The first model by Finnemore and Sikkink (Section 2.1) describes normative processes to capture the dynamics of international relations, whereas the models by Savarimuthu and Cranefield (Section 2.2), Hollander and Wu (Section 2.3), and Mahmoud *et al.* (Section 2.4) have been proposed in the research field of normative multi-agent systems. Since the identified individual processes that constitute all models are supported by relevant literature contributions, we provide an updated review of associated literature. The later three models represent incremental extensions of earlier models, and, in consequence, feature redundant elementary processes. In such cases, we refer the reader to the corresponding processes in earlier life cycle models.

2.1 Model 1: Finnemore & Sikkink

2.1.1 Overview

Norms have been traditionally studied in the social sciences [Crawford and Ostrom, 1995] (see also Finnemore and Sikkink [1998], Elster [1989], Bicchieri [2006]), but no consensus yet exists on how norms emerge and are subsequently adopted in a society. In order to understand the role that norms play in international politics, Finnemore and Sikkink [1998] introduced the concept of "life cycle" to model the origin and the dynamics of norms. They claimed that norms follow a specific pattern and that each portion of the life cycle is characterized by different actors and mechanisms. The term of life cycle was later imported and became particularly relevant for the study and modelling of normative multi-agent systems.

Finnemore and Sikkink's norm life cycle is a three-stage process, as shown in Figure 1: the first step is norm *emergence*, followed by norm *acceptance* (following Sunstein [1996], also called norm *cascade*), and the last stage is norm *internalization*. The move from norm emergence to norm cascade happens once the norm has been accepted by a certain amount of actors (the threshold point).

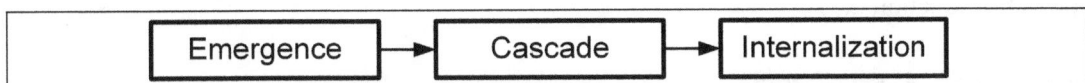

```
┌──────────────────────────────────────────────────────────┐
│  ┌────────────┐    ┌───────────┐    ┌─────────────────┐   │
│  │ Emergence  │───▶│  Cascade  │───▶│ Internalization │   │
│  └────────────┘    └───────────┘    └─────────────────┘   │
└──────────────────────────────────────────────────────────┘
```

Figure 1: Finnemore and Sikkink's Norm Life Cycle Model

It is important to mention that a norm does not necessarily complete a life cycle. If, for instance, a norm does not reach the threshold point, it will not move from the emergence stage to the cascade stage. The different stages of Finnemore and Sikkink's model are supported by examples coming from women's movement of suffragettes and laws of war.

2.1.2   Stage 1: Norm Emergence

At the origin of norms we find norm *entrepreneurs*, agents committed to persuade a critical mass to support new norms or to alter existing ones in order to achieve desirable behaviour in a state or community. As Hoffmann [2003] notes, leaders and entrepreneurs are not novel concepts in political science [Nadelman, 1990; Young, 1990; Schneider and Teske, 1992; Bianco and Bates, 1990]: "Entrepreneurship is a popular factor for explaining solutions to collective action problems, equilibrium choice, the emergence of cooperation as well as norms" (Hoffmann [2003], p. 8). As an example of a norm entrepreneur, Finnemore and Sikkink mention Henry Dunant, who played a crucial role in forming the norm that, in time of war, doctors and wounded soldiers should be treated as noncombatants and, by consequence, granted immunity.

The task of norm promoters is rarely easy. More often proposing a new norm implies competing with existing social contexts and established states of affairs. This means that one has to be ready to battle with competing norms or conflicting interests. The mechanisms by which individuals manage to convince other individuals is debated [Checkel, 1998; Risse and Sikkink, 1999]. Finnemore and Sikkink argue that the difficulty of the task explains why norm entrepreneurs frequently resumed to controversial or even illegal acts (such as the protests engaged by suffragettes, who refused to pay taxes and went on hunger strikes, among other things). Altruism, empathy and commitment to an ideal are the motives that Finnemore and Sikkink attribute to norm entrepreneurs to explain their dedication.

Observing norm emergence in international relations, Finnemore and Sikkink stress that norm entrepreneurs act within organizational platforms, like nongovernmental organisations. This facilitates the reaching of the threshold point and thus the emergence of the norm. In the context of international politics, empirical studies fix such threshold around one-third of the total states, even though some states are more critical to the adoption of a norm than others. The second stage (norm cascade) is reached when the threshold is passed.

Subsequent models, like Hollander and Wu [2011b], will refine Finnemore and Sikkink 's norm life cycle and will replace entrepreneurs by machine learning and cognitive approaches (Section 2.3).

2.1.3   Stage 2: Norm Cascade

We have seen that once the threshold of the critical mass is passed, according to the Finnemore and Sikkink's model, we move to the stage of norm cascade. This is called so because the acceptance rate of the new norm among the individuals increases rapidly. The mechanism that seems to govern the acceptance of a norm is *socialization*, a kind of persuasion by some agents to others to embrace a certain norm. In the case of states, such persuasion appears to lean against the need of a state to be recognized as a member of an international organisation. In other words, exactly as it happens to people, countries would be exposed to peer pressure. In particular, the desire to acquire or increment internal and international legitimation, the pressure of conformity and the need for norm leaders to increase their esteem seem to be the reason to respond to such a pressure.

2.1.4 Stage 3: Internalization

If a norm reaches the third and last stage, it becomes internalized. This means that such norm is acquired and not object of debate anymore. As Epstein stated, once a norm is accepted, people "conform without really thinking about it" (Epstein [2001], p.1). Examples of nowadays internalized norms are the abolition of slavery or the right to vote for women. But internalized norms can also be specific to certain professions. Finnemore and Sikkink mention the examples of doctors and soldiers, who become acquainted with different "normative biases": "Doctors are trained to value life above all else. Soldiers are trained to sacrifice life for certain strategic goals" (Finnemore and Sikkink [1998], p.905).

2.1.5 Discussion

Constructivists (to which Finnemore and Sikkink's approach belongs) have been criticized for failing to account how entrepreneurs hammer new norms or come to propose the alteration of existing ones, as well as how they manage to convince other critical agents in their vision. Hoffmann [2003] partially addresses such criticisms by building an agent-based model to explore the role of norm entrepreneurs. His model does not tackle the question of how entrepreneurs convince other agents, but focuses "on the unexamined assumption that a persuasive entrepreneur can influence the outcomes that arise from the interactions of heterogeneous, interdependent agents" (Hoffmann [2003], p. 13). His model shows that the constructivist's hypothesis of the role of norm entrepreneurs is indeed plausible. In particular, his aim is to understand under what conditions a norm entrepreneur can function as a norm catalyser for the emergence of new norms and the alteration of existing ones. Norm entrepreneurs turn out to be able to influence norm emergence even when they can reach only a small portion of the population (around 30%), and their influence increases with their reach. Hoffmann's model suffers (as the author himself acknowledges) from some limitations, like the assumption of a unique norm entrepreneur, the lack of communication among agents, agents' power is not modelled, and only non-complex norms are considered.

2.2   Model 2: Savarimuthu & Cranefield

2.2.1   Overview

The first life cycle model for norms we have encountered was proposed in the context of international relations. As we have seen, Finnemore and Sikkink [1998] directed their attention to human societies and to processes that can explain how norms emerge and spread within and among states. Ten years separate Finnemore and Sikkink's work from the second model we consider here, the life cycle model proposed by Savarimuthu and Cranefield [2009; 2011].

Savarimuthu and Cranefield's model comes from the study of simulation-based works on norms in the context of multi-agent systems. By looking at the various mechanisms employed by the researchers working on simulation on norms, they extend the three-stage model of Finnemore and Sikkink.

Savarimuthu and Cranefield's contribution came in two papers: the first one [Savarimuthu and Cranefield, 2009] presented a four phases norm life cycle (*norm creation, spreading, enforcement and emergence*), whereas the subsequent [Savarimuthu and Cranefield, 2011] included one additional stage (*identification*). For this reason, in the present section we will focus on the latter, more recent, contribution. For each step Savarimuthu and Cranefield provide a categorisation of the mechanisms that have been employed in the simulation-based works on norms, as shown in Figure 2.

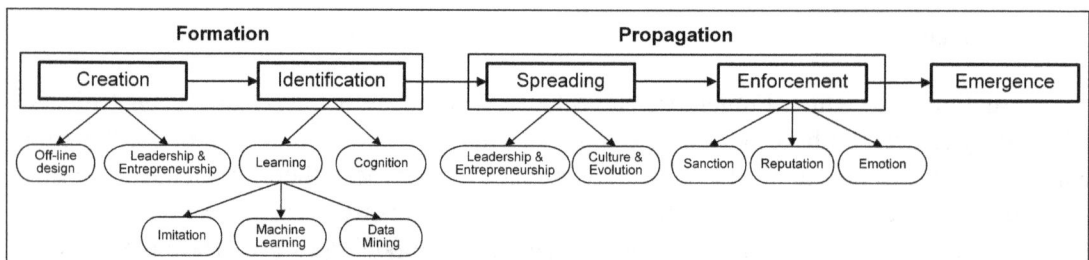

Figure 2: Savarimuthu and Cranefield's Norm Life Cycle Model

2.2.2   Norm Creation

Unlike Finnemore and Sikkink [1998], who acknowledged only the role of norm entrepreneurs for the creation of norms, Savarimuthu and Cranefield [2011] realize that in the context of multi-agent systems, norms can be created by three different approaches: off-line design, norm leaders and norm entrepreneurs. In off-line design the norm is introduced by an external designer and is hard-wired into the agents. Norm leaders, on the other hand, are powerful agents of the system that (following a democratic or an authoritarian process) create norms for the other agents to follow. Finally, norm entrepreneurs are not necessarily norm leaders. Similarly, as seen in Finnemore and Sikkink's model, an entrepreneur can propose a new norm that he thinks is beneficial to the society. But until the entrepreneur does not succeed to persuade the other agents to accept such norm, the norm is not a social norm.

Off-line Design One of the most well-known works in the area of off-line design is Shoham and Tennenholtz [1995]'s work on synthesising social laws, specifically in the traffic domain. In this specific context, off-line design implies that mobile robots (as traffic participants) are initialized with a set of traffic laws ('rules of the road') that have been computed at design time in order to prevent collisions at runtime. Such rules allow to minimize the need of a central coordinator on the one hand and that of a negotiation mechanism among agents on the other hand. Traffic laws provide the agents with a set of social laws that help them avoiding collisions. A multi-robot grid system is considered, where m robots can move on an $n \times n$ grid. Shoham and Tennenholtz suggest one can imagine rows and columns of that grid as lanes in a supermarket. In order to avoid the collision between robots (which happens when more than one robot occupies the same coordinate), some traffic laws are given. For instance, one may impose that in odd rows agents can move only right, in even rows they can move only left, in odd columns they can move only down and in even columns the only movement possible is up. Rules define also the priority when two or more robots approach a junction and how robots can change their movement direction Shoham and Tennenholtz [1995]'s work has subsequently been extended (e.g. to consider the minimality of social laws [Fitoussi and Tennenholtz, 2000]) and has found various adaptations in works on norm emergence (e.g. [Sen and Airiau, 2007; Mukherjee *et al.*, 2007]).

A similarly influential model from the sociological domain is Conte and Castelfranchi [1995b]'s evaluation of norms for the purpose of aggression control to facilitate cooperation in a stylized food-gathering society. In their model societies are selectively initialized as either *strategic* or *normative*, where strategic agents systematically attack fellow food-carrying agents, whereas normative ones accept a notion of possession, thus promoting a higher level of survival at the macro level. Results have shown that normative populations do better than strategic ones. However, in mixed populations strategic agents do much better than the normatives. The reason is that non-normative agents benefit from the behaviour of normatives.

Castelfranchi *et al.* [1998] have further extended the model to consider the role of reputation (see Section 2.2.5). The role of reputation is considered also in Hales [2002], which extended Castelfranchi *et al.*'s food-consumption problem by assigning agents to the group of normative agents or to the group of cheaters.

Walker and Wooldridge [1995] observe that the simplicity of off-line design models comes at a price. To be truly beneficial, such approach requires that all characteristics of a system should be known a priori (which is not the case for open systems, for example). Another difficulty is that it is extremely costly and time-consuming to constantly reprogram agents, which is required in case agents' goals change, as it happens in complex systems. Moreover, Savarimuthu and Cranefield [2011] note that it is not realistic to assume that all agents follow a given norm.

Leadership and Entrepreneurship Mechanisms   Leaders are agents who have the social power and abilities to persuade other agents to accept a norm. Leadership mechanisms have been employed for norm emergence and norm spreading (see Section 2.2.4).   Verhagen [2001] considers agents with a certain degree of autonomy and a normative advisor (as in Boman [1999]'s approach) from whom they receive comments on an agent's decision to follow or not to follow a norm. Once an agent decides to follow a specific norm, it announces it to the whole society. The normative advisor as well as other agents can send feedback to that agent, who may assign a greater weight to the comments received from the leader.

In Savarimuthu *et al.* [2008a] a society can have several normative advisors (or role models) who give advice to agents who are their followers. Agents are connected to each other through one social network topology among fully connected networks, random networks and scale-free networks. The interesting twist is that an agent can be at the same time a role model for some agents and a follower of some other agent. Since several norm leaders can exist, different norms can emerge in the society.

Norm entrepreneurs were notably introduced in Finnemore and Sikkink's norm life cycle model, presented in Section 2.1. Hoffmann [2003] has experimented on the notion of norm entrepreneurs, as seen in the Discussion subsection of Section 2.1.

2.2.3   Norm Identification

The first norm life cycle model proposed by Savarimuthu and Cranefield [2009] consisted of four stages (norm creation, spreading, enforcement and emergence). The idea being that, as in [Finnemore and Sikkink, 1998], once a norm is created, it may spread in a society if certain conditions are satisfied. However, in [Savarimuthu and Cranefield, 2011], they added the *identification* step between norm creation and spreading. Such step is needed in all those situations in which a norm has not been explicitly created, for example when a norm results from the interaction process among agents. In those cases, agents have first to be able to identify the created norms. Simulation-based works on norms have explored two approaches for norm identification: agents can learn new norms by imitation, machine learning or data mining mechanisms; alternatively, agents can use their cognitive abilities to infer and recognize the norms of a system.

Learning Mechanisms – Imitation   Among the simulation models that experimented on learning mechanisms based on imitation is that of Epstein [2001]. Using a driving setting in which agents can observe whether other agents (within a certain radius) drive on the right or on the left, Epstein showed that agents conform to the driving preferences of the majority of the observed agents. Imitation mechanisms can explain the identification and the spreading of a norm.

Yet, some authors, like López y López and Márquez [2004] as well as Campenni *et al.* [2009], cast some doubts on the claim that such mechanisms can explain the co-existence of different norms in a group of agents. Instead of seeing norms are hard-wired in the agents, Campenni *et al.* [2009] imagine the interaction between agents coming from different societies. Their goal is to investigate the role of cognition in norm recognition: How do agents tell that something is a norm? In their model, there are four scenarios, some actions that are context-specific and one action that is common to all scenarios. In one set of simulations, agents can change contexts, whereas in another set of simulations, at a certain moment, agents must stay in the context they have reached and can interact only with agents that are in the same context (imagine a situation in which a population is split into two groups and each group is constrained to not have contacts with the other group). The purpose of this second set of simulations is to show that frequency may be a sufficient (but not necessary) condition for agents to converge to the same action. Results show that new norms can emerge, eventually giving rise to the competition between two rival norms.

Learning Mechanisms – Machine Learning   Shoham and Tennenholtz [1992a] employed co-learning, a reinforcement learning mechanism that makes an agent choose the strategy that revealed to be the most successful in the past. They showed that norm emergence decreases with the decrease of the frequency of the updates of an agent's strategy. The efficiency of norm emergence turned out to decrease also with the increase of an agent's memory flush.

Building on the scenario introduced in [Conte and Castelfranchi, 1995b], Walker and Wooldridge [1995] ran 16 experiments with different parameters for the size of the majority and the update function (the latter could depend on the majority rule, on the memory flush or on communication mechanisms). Results showed that the network topology and com-munication may play an important role and, hence, more simulations are needed to better understand mechanisms for norm emergence.

More recently, norm emergence has been investigated using social learning in a model in which agents repeatedly interact with other agents by Sen and Airiau [2007]. Experiments took into account different population sizes, various learning strategies, and number of available actions. The specific situation is that of learning of which side of the road to drive on but also the problem of who has the priority if two agents gain a junction at the same time. The outcomes confirm that such a mode of learning is a robust mechanism for the emergence of social norms.

Learning Mechanisms – Data Mining   An approach to norm identification that uses association rule mining to identify obligation norms is Savarimuthu *et al.* [2010b]'s *Obligation Norm Inference* (ONI) algorithm. Such model enables agents to sense their environment, memorize experiences and observations as well as normative signals, which build the basis for the identification of personal norms (p-norms) and group norms (g-norms). The memorized event episodes are then mined for obligation norms using association rules algorithms. The agent-based simulation experiment considers a virtual restaurant in which agents may not know whether the restaurant expects the customers to order and pay for the food at the counter before eating or if they are expected to order, consume the food and pay only before leaving. Another protocol agents may need to identify is the tipping norm: in some countries, for example, tipping is expected (in the USA, for instance), whereas in others (like most countries in Europe) it is not expected. The difficulty in identifying an obligation norm is that a sanction is triggered by the absence of an action (a customer in a restaurant may be sanctioned if he is not tipping the waiter). Savarimuthu *et al.* [2013a] propose a corresponding approach for the identification of prohibition norms.

Savarimuthu and Cranefield [2011] observe that data mining is a promising approach. However, explicit signals for sanctions or reward have to be present in order for norms to be easily identified.

Cognition   The EMIL-A architecture [Andrighetto *et al.*, 2007; Campenni *et al.*, 2009; Andrighetto *et al.*, 2010][1] is a cognitive architecture to explore how agents' mental abilities may explain the acquisition of new norms. Reinforced candidate norms are identified from observed normative information (represented as normative frame) that traverses different memory layers, representing the transition from short-term experiences to long-term memory. Once established, normative beliefs are held in a *Normative Board*, along with associated normative action plans. These internalized normative beliefs inform the agent's goal generation, decision-making and action planning. The previously discussed work by Savarimuthu *et al.* [2010b] also proposed an architecture for agents to identify norms using agents' cognition abilities.

---

[1]Campenni *et al.* [2009]'s contribution is a notable extension of Andrighetto *et al.* [2010]'s work.

2.2.4 Norm Spreading

Once a norm has been explicitly created or agents have identified it, the norm can start being spread in the society. Among the different mechanisms that can serve this purpose, there are leadership and entrepreneurship that we already encountered in the norm creation stage, but also cultural and evolutionary mechanisms.

Culture and Evolution   Cultural and evolutionary mechanisms have been considered in [Boyd and Richerson, 1985; Chalub *et al.*, 2006]. According to Boyd and Richerson [1985] social norms can be propagated along three types of transmissions: vertical, horizontal and oblique. *Vertical relationships* describe the intergenerational transmission of norms by parents to offspring, whereas *horizontal transmission* occurs among peers of a given generation. *Oblique relationships* combine the former two and describe the unidirectional dissemination of norms by authority figures towards their contemporary subalterns. Vertical relationships are constrained to the intergenerational sharing of norms which makes them particularly applicable to evolutionary models such as Axelrod's norm game [Axelrod, 1986]. Horizontal approaches assume a uniform social structure, which limits this approach to abstract group or society representations, as is the case for large parts of the norm emergence work (e.g. [Sen and Airiau, 2007; Villatoro *et al.*, 2011a; Mihaylov *et al.*, 2014; Airiau *et al.*, 2014]; Section 4). The last relationship type lends itself well to model inter- and intra-generational norm transmission for comprehensive society representations that consider power and authority structures. Examples for this include Franks *et al.* [2014]'s use of *Influencer Agents* to drive the norm convergence, or Yu *et al.* [2015]'s hierarchical approach to information sharing.

Savarimuthu and Cranefield [2011] note that if cultural and evolutionary mechanisms can explain how a norm is spread, they cannot answer the question of how a norm is internalized in the first place.

2.2.5 Norm Enforcement

The existence of a norm presupposes that such norm can be violated. Norm enforcement mechanisms serve to deter agents from violating a norm. This can be done through punishment, via some mechanisms that negatively affect the reputation of a norm violator, or again by affecting the agent's emotions (for example, by instilling a sense of guilt in the norm violator). Savarimuthu and Cranefield [2011] stress that norm enforcement can also play a role in the spreading process of a norm. Observing the punishment of a norm violator can either discourage other agents from violating that norm or identifying that norm, in case it was not explicitly created.

Sanctions   The most well-known work on external sanctions is Axelrod [1986]'s norm game that specifically explores the notion of metanorms, i.e. the sanctioning of non-sanctioning observers of violations, to sustain a society's norm.[2]   An essential challenge of normative regulation (in artificial systems as in real life) is the balance of cost and effect of sanctions, both to minimize the cost of enforcement, while maximizing the effect in order to regulate behaviour effectively [Axelrod, 1986; Horne, 2001; Savarimuthu *et al.*, 2008a].   Mahmoud *et al.* [2012; 2015] refine Axelrod's model by investigating the effect of dynamic punishment, and ultimately propose an alternative to Axelrod's evolutionary approach based on individual learning to produce a model in which norms can stabilize within a given generation.

In López y López [2002; 2003] a model where agents have goals and different personalities is developed. Punishments and rewards are considered only when they affect an agent's goals.

Reputation   A positive or negative opinion about one agent from the interacting agents in a society can play a substantial role in the norm compliance in a group of agents.

In Castelfranchi *et al.* [1998]'s and Younger [2004]'s models, ostracism is an implicit result of reputation sharing, which leads to the exclusion of individuals from future interaction. In particular, Castelfranchi *et al.* [1998]'s game reconsiders Conte and Castelfranchi [1995b]'s stylized food-gathering society seen in Section 2.2.2, with the addition of normative reputation: agents learn the reputation of other agents, that is, they learn whether an agent is normative or strategic (i.e. a cheater). However, in order to be profitable, the information about cheaters must be communicated to other agents. In the context of multi-agent systems Perreau de Pinninck *et al.* [2010] propose a distributed mechanism that affords the isolation of violating nodes in the context of peer-to-peer applications. They evaluate its properties for various network topologies.

Emotion   Staller and Petta [2001] introduce an extension of the cognitive agent architecture JAM [Huber, 1999] with components to augment the rational agent model with emotion appraisal processes, an aspect considered essential to mediate any form of norm enforcement [Scheve *et al.*, 2006].   Fix *et al.* [2006] propose a model of normative agents that include the display of emotional responses to normative actions. In this work the agents' internal states are represented using reference nets [Valk, 1998], a variant of Petri nets.

---

[2]Axelrod's contribution was impressive and extremely influential. However, it should be noted that Galan and Izquierdo [2005] have shown that his results are not stable. When running many more simulations of Axelrod's model and for longer, opposite results can be obtained. As the authors also stress, one should not forget that their analysis required computational power which was not available when Axelrod proposed his model.

2.2.6 Norm Emergence

Once a norm has spread across a certain proportion of the society (according to different simulation results, the minimum required is a third of the population), it is said that the norm has emerged. This implies that a significant proportion of the population recognizes and follows that norm. It is worth noticing, however, that such process can be reverted. A norm may lose its appeal in a group and is hence either abandoned, replaced or modified by a competing one.

No specific category of empirical work on norms is associated with norm emergence. However, there is one category whose impact is notable across all stages of norm development. This is the consideration of network topology, as described in the Transmission part in Section 2.3.2.

2.2.7 Discussion

Savarimuthu and Cranefield [2011]'s life cycle model is an extension of the life cycle introduced by Finnemore and Sikkink in [Finnemore and Sikkink, 1998]. There are, however, two main differences.

The first one is that, whereas Finnemore and Sikkink's model was thought for human societies, Savarimuthu and Cranefield direct their attention to normative multi-agent systems and to simulation studies of norms using software agents. The second difference is that Savarimuthu and Cranefield not only capture two additional steps in their model, but also that for each phase, they consider more mechanisms.

2.3 Model 3: Hollander & Wu

2.3.1 Overview

To date, the most complex norm life cycle model has been proposed by Hollander and Wu [2011b]. Their model refines the ones initially introduced by Finnemore and Sikkink [1998] (Section 2.1) and Savarimuthu and Cranefield [2011] (Section 2.2), resulting in a total of ten *normative processes*, namely *creation, transmission, recognition, enforcement, acceptance, modification, internalization, emergence, forgetting*, and *evolution*. In contrast to the earlier models, Hollander and Wu identify three superprocesses (*enforcement, internalization*, and *emergence*) that combine elementary processes and characterize their high-level function. Note that the superprocess labels are borrowed from the most essential elementary process out of all processes they combine. A further novelty is the interpretation of *emergence* as an iterative process, and *evolution* as a metaprocess the authors refer to as "end-to-end process" [Hollander and Wu, 2011b]. The schema in Figure 3 provides a systematic overview of the complete life cycle. Where existing, the superprocesses are represented as boxes comprising their elementary processes, with the corresponding superprocess label highlighted in bold font. We will briefly outline the entire life cycle before introducing the individual processes in greater detail.

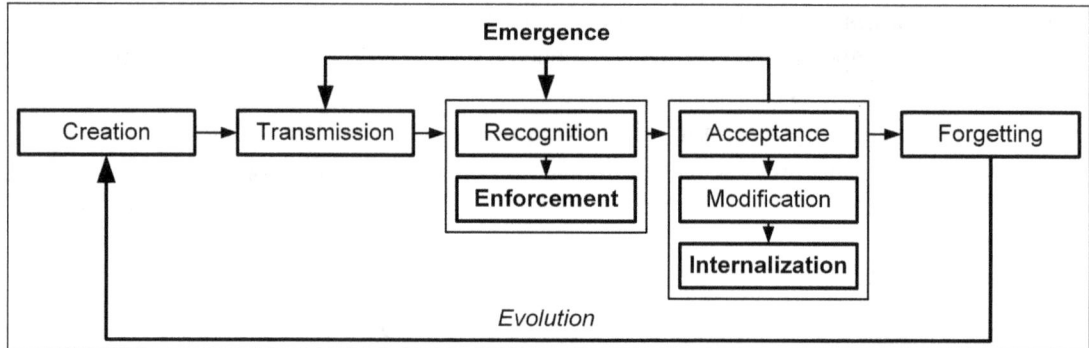

Figure 3: Hollander and Wu's Norm Life Cycle Model

Initially, potential norms are explicitly created, before being transmitted to the wider society, and rely on recognition and enforcement processes (captured in the superprocess *enforcement*) to promote their adoption. The superprocess *internalization* involves the decision whether to accept a norm, potentially modifying it, and finally, internalizing it, and thus becoming an enforcer of the norm itself. The subsequent cyclic reinforcement of the norm, including transmission, enforcement and internalization (tagged *emergence*), determines whether the initial *potential norm* becomes a norm. If attaining normative status, norms undergo a continuous refinement that requires reiteration through the elementary processes to gain salience. Any norm modification, such as the adaptation to new circumstances, implies that some normative content is forgotten. Swipe-card payments for bus services, for example, make it increasingly permissible for individuals to enter buses through arbitrary doors, instead of requiring the traditional entry through specific doors for payment. Contrasting the gradual forgetting of normative content, norms can be superseded by alternative norms, in which case the original norm is forgotten in its entirety. For example, over the past decades in many Western countries the general tolerance towards smoking in public places has been progressively replaced with general rejection.

In the following, we will discuss selected processes in greater detail and contextualize those with the earlier life cycle models as well as recent developments.

2.3.2 Life Cycle Processes

Creation   Similar to Savarimuthu and Cranefield [2011], Hollander and Wu acknowledge that norm creation involves a wide range of different processes, including methods found in the natural world [Boella *et al.*, 2008; Finnemore and Sikkink, 1998; López y López *et al.*, 2007; Savarimuthu and Cranefield, 2009], such as spontaneous emergence from social interaction, decree by an agent in power, or negotiation within a group of agents. However, in the context of work on NorMAS, Hollander and Wu identify two primary methods of norm creation, namely off-line design [Conte and Castelfranchi, 1995a; Shoham and Tennenholtz, 1995] and autonomous innovation [Hollander and Wu, 2011b]. While off-line design assumes that experimenters create the norms a priori and inject those into instantiated agents, autonomous innovation (akin to 'on-line design') assigns the role of norm creation to agents themselves.

Notable works in the area of off-line design include Shoham and Tennenholtz [1995] and Conte and Castelfranchi [1995b], as already discussed in Section 2.2.2.

Autonomous innovation covers a broader range of approaches, ranging from the adoption of specific strategies to the challenging problem of ideation, namely giving agents the ability to produce novel ideas without external input.

In contrast to previous models' norm creation mechanisms in the form of norm leadership/entrepreneurship (see Sections 2.1 and 2.2), Hollander and Wu [2011b] identify two types of mechanisms used for autonomous innovation, namely:

- Game-theoretical and machine learning approaches (e.g. Sen and Airiau [2007], Mukherjee *et al.* [2007], Perreau de Pinninck *et al.* [2008], Urbano *et al.* [2009], Sen and Sen [2010], Savarimuthu *et al.* [2010b]), and

- Cognitive approaches (e.g. Savarimuthu *et al.* [2010b], Andrighetto *et al.* [2007]).

Even though many models use a combination of those mechanisms,[3] their application tends to serve distinctive purposes. Game-theoretical approaches emphasize the identification of optimal strategies from a set of given strategies, thus representing an incremental step from off-line design towards autonomous norm innovation. Machine learning is generally used in conjunction with game-theoretical approaches, mostly to represent a notion of memory (e.g. Sen and Airiau [2007], Mukherjee *et al.* [2007]).

Essential work that combines game-theoretical and machine learning approaches is the research field of *norm emergence* or *convention emergence*. This field concentrates on the identification of factors that promote high convergence levels for norms within the observed society. While decision-making itself is modelled as some form of game (with 'rules of the road' [Shoham and Tennenholtz, 1995] as the preferred coordination game), agent components such as memory are represented using machine learning (commonly reinforcement

---

[3]Examples for combining game-theoretical and machine learning approaches are provided by Sen and Airiau [2007] and Mukherjee *et al.* [2007]; an example for the combined use of machine learning and cognitive approaches is Savarimuthu *et al.* [2010b]'s work.

learning in the form of Q-learning [Watkins and Dayan, 1992]). Depending on the aspect of interest, the model is augmented with additional mechanisms to investigate the influence of memory (e.g. Villatoro *et al.* [2009]), characteristics of network topologies and structural dynamics (e.g. Savarimuthu *et al.* [2007], Villatoro *et al.* [2009], Sen and Sen [2010], Villatoro *et al.* [2013]), norm transmission mediated by social learning (e.g. Sen and Airiau [2007], Mukherjee *et al.* [2007; 2008], Airiau *et al.* [2014]), as well as adaptive sanctioning (e.g. Mahmoud *et al.* [2012; 2015]).

Sen and Airiau [2007], for example, let agents engage in social interaction in the context of the 'rules of the road' scenario (described in Section 2.2), in which cars approach an unregulated intersection and have to identify an optimal coordination mechanism, such as 'yield to the right', and prevent deadlocks (both cars yield) or collision.[4] Agents memorize past encounters and adjust their behaviour based on the success of their action. As part of their evaluation, Sen and Airiau explore different population sizes, action spaces and learning algorithms to show how agent societies can autonomously arrive at stable norms.

Further approaches investigate the influence of hierarchical structures on the distribution of norms (e.g. Franks *et al.* [2013; 2014], Yu *et al.* [2013; 2015]).[5]

While work in the area of norm emergence concentrates on the interactions and corresponding macro-level outcomes, cognitive approaches concentrate on the mechanics of normative agent architectures. Cognitive norm architectures contextualize perceived behaviour with existing beliefs to infer normative content and/or consider normative beliefs in their deliberation process. Approaches of this kind generally consider more complex forms of learning. They further invoke semantically rich norm representations and processes that come closest to what we can describe as *ideation* [Ehrlich and Levin, 2005], i.e. proposing behaviours that potentially qualify as normative, and selectively filtering those.

Representative works that apply cognitive approaches include the *Beliefs-Obligations-Intentions-Desires* (BOID) architecture [Broersen *et al.*, 2001; Broersen *et al.*, 2002] which extends the widely adopted Belief-Desire-Intention (BDI) architecture [Bratman, 1987; Rao and Georgeff, 1995] with an obligation component that preempts the goal generation and prioritizes the individuals' obligations. In this approach, obligations are statically embedded in an agent's belief base.

While BOID emphasizes normative reasoning, alternative approaches propose mechanisms to facilitate norm identification and decision-making, along with the involved micro-/macro-level interaction, as in the cognitive architecture EMIL [Andrighetto *et al.*, 2007; Campenni *et al.*, 2009; Andrighetto *et al.*, 2010], that extends the BDI concept with the ability to acquire new norms, which we discussed in Section 2.2.3.

Cognitive approaches such as Savarimuthu *et al.* 's norm identification frameworks for

---

[4]We will come back to this scenario in greater detail in Section 4, given of its relevance in the area of norm synthesis.

[5]We will discuss the field of norm emergence in more detail in Section 4.

obligation [Savarimuthu *et al.*, 2010b] and prohibition norms [Savarimuthu *et al.*, 2013a] rely on notions of machine learning to afford realistic agent representations [Savarimuthu *et al.*, 2011; Ossowski, 2013]. Further examples for the combined use of cognitive and machine learning components include the identification of normative content from action and/or event sequences (e.g. Savarimuthu *et al.* [2010a]), the implementation of alternative learning mechanisms beyond experiential learning or 'learning by doing', such as social/observational learning [Bandura, 1977] as applied by Epstein [2001], Hoffmann [2003], as well as Sen and Airiau [2007]. Another combined use of cognitive and machine learning is to facilitate the use of direct communication (e.g. used by Verhagen [2001] as well as Walker and Wooldridge [1995]).

Transmission   The norm transmission process in Hollander and Wu's model (equivalent to the spreading process in Savarimuthu and Cranefield [2011]'s model), considers three components that characterize how information is spread. Those include:

- the nature of *Agent Relationships*,

- the applied *Transmission Techniques*, and

- the underlying *Network Structure*.

Agent Relationships   Similar to Savarimuthu and Cranefield [2011], Hollander and Wu share Boyd and Richerson [1985]'s observation of relationship types as either being vertical, horizontal or oblique, an aspect we discussed in the context of Savarimuthu and Cranefield's model (Section 2.2.4).

Transmission Techniques   Beyond the identification of relationships, Hollander and Wu [2011b] identify two transmission techniques for norms, the first being *active transmission* in which norms are actively broadcast throughout the relationship networks. Alternatively, agents can use *passive transmission* and absorb perceived normative information. Examples of mechanisms to facilitate active transmission include direct communication, whereas observation of the social environment (on the part of a norm recipient) is an example of passive transmission.

In most simulation works, active transmission is used to convey normative content by direct communication or in the form of sanctions. Examples include Hoffmann [2005], who uses proactively communicating norm entrepreneurs to promote convergence, as well as the previously mentioned work by Franks *et al.* [2013], or Yu *et al.* [2010; 2015]'s use of *supervisors* to model hierarchical communication between networked multi-agent systems. Further examples from the sociological domain include Castelfranchi *et al.* [1998]'s and Younger [2004]'s society models that rely on reputation sharing for the purpose of promoting cooperation.

Examples of passive communication are used to represent notions of imitation or social learning. Examples include Verhagen [2001]'s work on norms learning, as well as the work on the impact of social learning on norm convergence (e.g. Nakamaru and Levin [2004], Sen and Airiau [2007], Airiau *et al.* [2014]) and synthesis (e.g. Frantz *et al.* [2015]). An example of the use of passive transmission in social scenarios is Flentge *et al.* [2001]'s representation of imitation by copying memes from successful neighbours.

Network Structure   The third aspect of norm transmission is the nature of the underlying connectivity structure that acts as an information transport medium. Depending on the objective, the connectivity structure is conceived as a multi-dimensional grid environment or as network topology of varying complexity.

In grid environments, agents are stationary or mobile, and observe agents within their specified neighbourhoods, and can, depending on their neighbourhood configuration, perceive adjacent cells. Agents' grid environments are generally modelled as von Neumann neighbourhoods – in which agents can sense orthogonally adjacent cells – or Moore neighbourhoods – in which agents can sense all adjacent cells.

The modelling of norm transmission via network structures permits the configuration of more complex relationship networks, with network topologies of equal degrees of connectedness (e.g. as fully connected networks), as well as random connectivity (random networks [Erdős and Rényi, 1959]). Alternatively, networks can display varying degrees of connectedness, such as small world networks [Watts and Strogatz, 1998] that simulate sparse links between communities characterized by dense internal connectedness. Scale-free network topologies [Barabási and Albert, 1999] work on the far end of the spectrum and produce a structure characterized by power law distributions, with individuals being centred around densely-connected hubs.

In analogy to the stationary or mobile configuration in a grid environment, a further important aspect is whether network topologies are static or dynamic at runtime. Effects of complex network topologies on norm emergence have been explored by Zhang and Leezer [2009], Franks *et al.* [2014], and Sen and Sen [2010]. Villatoro *et al.* [2009] put specific emphasis on the interaction between memory size and the chosen topology, whereas Airiau *et al.* [2014] concentrate on the effect of social learning across different topologies. Savarimuthu *et al.* [2007] and Villatoro *et al.* [2011a; 2013] explore the effect of dynamic topologies on norm emergence.

Recognition   In Hollander and Wu's model, the processes *creation* and *transmission* are followed by the superprocess *enforcement* that consists of the subprocesses *recognition* and *enforcement* (see Figure 3). Norm recognition is similar to Savarimuthu and Cranefield's account of *norm identification* and describes the agent's ability to recognize the norms enacted in the observed society or group. Means to do so include communication with norm participants (as is the case with human societies [Henderson, 2005]) as well as observational learning. Similar to technological approaches in the context of norm creation, earlier models relied on off-line identification of agents as norm followers and deviants (e.g. Castelfranchi *et al.* [1998], Hales [2002]), whereas recent models apply more sophisticated mechanisms to identify norms, which include machine learning [Sen and Airiau, 2007; Mukherjee *et al.*, 2007; Savarimuthu *et al.*, 2013b; Frantz *et al.*, 2015] and/or cognitive approaches [Savarimuthu *et al.*, 2010b; Andrighetto *et al.*, 2007]. Since the recognition of norms may involve the observation of sanctions, it is closely related to enforcement.

Enforcement Norm enforcement describes the application of sanctions to stimulate adherence to the normative content. Sanctions can be positive (in the form of rewards) or negative in nature and can further be differentiated by their source, that is whether they originate from internal or external sources.

For this purpose Hollander and Wu differentiate three types of enforcements:

- Externally Directed Enforcement

- Internally Directed Enforcement

- Motivational Enforcement

Externally Directed Enforcement Externally directed enforcement describes sanctioning by an outside observer that witnesses and reacts to a norm violation or an agent's refusal to accept a transmitted norm (e.g. a follower rejecting a leader's imposed norm) [Flentge *et al.*, 2001; Galan and Izquierdo, 2005; Savarimuthu *et al.*, 2008b].

Applied sanctions can be of economic nature (e.g. reducing or limiting access to resources), affect the violator's reputation (e.g. shunning, ostracism) [Axelrod, 1986; Castelfranchi *et al.*, 1998; Hales, 2002; Younger, 2004] (as seen in Section 2.2.5), or prevent it from propagating deviance to others (e.g. by preventing procreation in the case of vertical norm transmission [Flentge *et al.*, 2001]) [Caldas and Coelho, 1999].

The prototypical example for external sanctions is Axelrod's norm game [Axelrod, 1986], as discussed in Section 2.2.5 in the context of Savarimuthu and Cranefield's life cycle model.

Internally Directed Enforcement Sanctions of internal origin rely on an individual's self-enforcement triggered by the violation of internalized norms. The prototypical mechanism for internally motivated norm enforcement is the activation of emotions (discussed in greater detail in Section 2.2.5).

Motivational Enforcement   Hollander and Wu further identify the notion of motivational enforcement, which is essentially a special case of internally directed enforcement. It describes the implicit commitment of all individuals to follow system-wide norms if they are aligned with an individual best interest, an aspect understood as conventions [Lewis, 1969]. A classical example is the convention of uniform road side use: the precise strategy (i.e. whether to drive on the left or right side) is secondary to the complete acceptance and internalization by the society since unilateral deviation produces suboptimal outcomes (i.e. accidents caused by ghost drivers).

Internalization   Processes that are essential for norm emergence in Hollander and Wu's model are associated with the superprocess norm internalization. Hollander and Wu differentiate between *Acceptance*, *Modification*, and *Internalization* (as the terminating subprocess of the superprocess *Internalization*).

The acceptance of enforced norms is the starting point for the internalization of norms by individuals and decisive for the emergence of norms, since individuals either decide to accept or reject socially imposed norms based on the compatibility with their personal beliefs, desires and intentions. Possible outcomes are the acceptance of a new norm, the substitution of an existing conflicting norm, or its rejection. Acceptance is operationalized as some form of cost-benefit analysis [Meneguzzi and Luck, 2009].

If agents decide to accept norms, their integration into the internal cognitive structures requires the transformation of norms from an objectified outside perspective to a subjective representation that involves an individual's biases, inaccuracies of perception, etc. This potentially leads to a modified understanding of that norm, an aspect that affects the norm during its further progression in the life cycle.

Finally, the accepted and potentially modified norm is internalized by the receiving agent. Compared to the other stages of the norm life cycle, this process has found limited explicit attention. In most applications, individuals simply adopt the accepted norms without further refinement or adaptation. From a motivational perspective, this is compatible with measures that suggest that the absence of external pressures is indicative of complete norm internalization [Epstein, 2001]. However, this view only accounts for subsequent norm adherence, but cannot explain violations further down the track. Refined approaches evaluate the effect of the internalized norm and on the decision-making process. An important example is Verhagen [2001]'s work, in which agents seek increasing alignment with their associated group by comparing and internalizing corresponding action probabilities. Alternatively, as done in the BOID architecture [Broersen *et al.*, 2001], internalized norms can be maintained separately from personal strategies and activated selectively depending on situation-specific autonomy values [Broersen *et al.*, 2002].

In their original survey, Hollander and Wu [2011b] highlighted the limited explicit focus on internalization, especially in comparison to life cycle processes such as enforcement. However, recent works in the area of NorMAS reveal more explicit treatments of internalization, generally in the form of continuous probabilistic adaptation of strategy choices based on reinforcement learning (e.g. Salazar *et al.* [2010], Villatoro *et al.* [2013], Franks *et al.* [2014], Airiau *et al.* [2014], Frantz *et al.* [2014b; 2015], Yu *et al.* [2015]), or by using thresholds for the adoption of new strategies (e.g. Hollander and Wu [2011a], Mihaylov *et al.* [2014]). In Section 2.6 we provide a comprehensive overview of internalization mechanisms used in works on normative multi-agent systems.

Emergence   In contrast to all earlier models, Hollander and Wu conceive emergence as a dynamic macro-level process that describes a cyclic iteration involving the transmission of the internalized norm to new participants. This is followed by enforcement (based on the subprocesses *Recognition* and *Enforcement*) to drive the internalization (composed of subprocesses *Acceptance*, potential *Modification*, and *Internalization*) of the norm by new subjects, who themselves participate in the spreading of the norm – ultimately leading to the norm's emergence as a macro-level phenomenon. This emergence understanding is aligned with Savarimuthu and Cranefield's, who interpret emergence as the final stage of the norm life cycle, but do not explicitly reflect the cyclic reinforcement of norms by reiterating through the formation stage. Finnemore and Sikkink's life cycle model maintains a different emergence interpretation and associates emergence with the micro-level creation of a norm, e.g. via entrepreneurship, before sharing and penetrating the wider society.

The exploration of emergence characteristics is strongly tied to the applied modelling technique. Game-theoretical approaches evaluate emergence by identifying stabilising strategy choices (equilibria) chosen from a set of given alternative strategies. The dominant strategy choice is then interpreted as the emergent norm (see e.g. Axelrod [1986], Mukherjee *et al.* [2007], Zhang and Leezer [2009]). Since agents are represented as structurally uniform selfish rationalizers with a minimal action repertoire, the exploration is focused on macro-level outcomes. Cognitive approaches, on the other hand, do permit a macro-level observation of specific norms, but furthermore, allow a more realistic reconstruction of micro-level processes. This includes detail and diversity of individuals' cognitive structures, the precise level and nature of enforcement (see e.g. Caldas and Coelho [1999], Savarimuthu *et al.* [2008b]), the use of richer norm representations, diverse action sets, and a variety of norm learning mechanisms (e.g. based on experiential learning, social learning and direct communication) [Savarimuthu *et al.*, 2011].

Models can further address infrastructural aspects, such as the impact of different connectivity structures on normative outcomes. Related findings suggest that scenarios in which normative behaviour is transmitted from neighbours (e.g. in grid environments) tend to result in the dominance of a single norm, whereas individualized learning promotes the emergence of diverse normative configurations [Boyd and Richerson, 1985; Boyd and Richerson, 2005; Nakamaru and Levin, 2004]. While the application of network structures can lead to stronger normative diversity, experimental results suggest that the impact of the actual network topology is secondary to its dynamic nature (as opposed to static networks) [Bravo *et al.*, 2012]. However, the convergence of conventions (and emergence of local subconventions) can be controlled by maintaining links to distant nodes [Villatoro *et al.*, 2009].

Forgetting & Evolution   In contrast to the earlier models by Finnemore and Sikkink as well as Savarimuthu and Cranefield, Hollander and Wu are the first to complete the norm life cycle by explicitly considering the process of *Forgetting*. In this conception forgetting is essential to sponsor the evolutionary refinement of norms, since continuously changing norm contexts may render existing norms irrelevant. An example is the normalized use of smart devices in education, with proactive integration of social media platforms such as Facebook into the learning environment. This is in opposition, or at least in competition, to traditional norms that ban the use of mobile devices in classroom environments. Once forgotten, norms make space for new norms that are better adapted to environmental needs, which constitutes the end-to-end process that closes the evolutionary loop of the norm life cycle.

2.3.3   Discussion

As mentioned at the outset of this section, this model proposed by Hollander and Wu introduces the to date most comprehensive life cycle model. The model not only considers abstract high-level processes (superprocesses), but decomposes those into elementary processes that capture large parts of contemporary research and, beyond this, identify gaps in normative agent architectures (such as the explicit consideration of *Norm Acceptance*) to produce more comprehensive representations of human reasoning processes. In addition to the fine-grained nature, this model further deviates from the linear operation of previous models by identifying emergence as a metaprocess that links individual processes and results in a continuous iteration through elementary processes. Beyond the 'completion' of the life cycle by considering the abandoning of norms, a further essential novelty is the consideration of norm evolution as a continuous process that affords both the modification and the substitution of norms over time.

2.4   Model 4: Mahmoud et al.

Overview   The latest life cycle model has been proposed by Mahmoud *et al.* [2014b]. Similar to the earlier life cycle models developed in the context of NorMAS, their work is based on a comprehensive literature review, both considering individual works as well as previous life cycle models. In contrast to Hollander and Wu's detailed model, their approach identifies five core processes (*Creation, Emergence, Assimilation, Internalization, Removal*) with a further decomposition of selected processes as shown in Figure 4. Since this model has only been briefly described by the original authors themselves and strongly builds on concepts introduced in the context of Hollander and Wu's earlier, more detailed model, we provide a concise overview at this stage, before discussing the novel contributions in more detail.

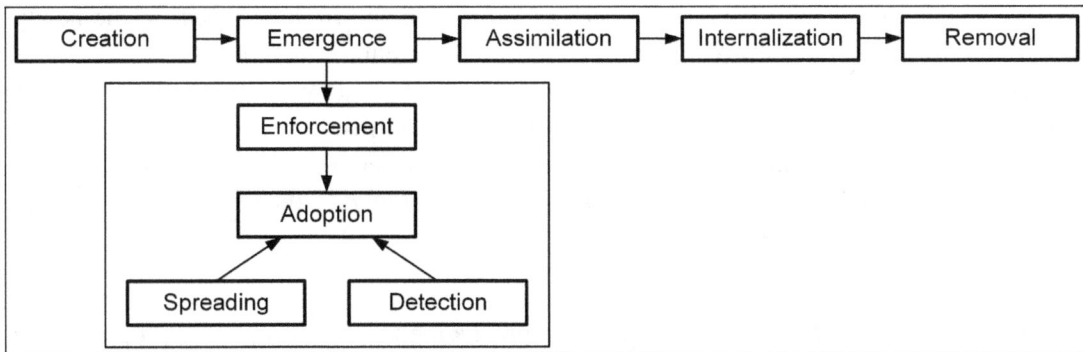

Figure 4: Mahmoud et al.'s Norm Life Cycle Model

Processes   The initial process, as with most other life cycle models is *Creation*, which operates based on mechanisms described by Savarimuthu and Cranefield [2011], namely off-line design, autonomous innovation and social power (see Section 2.2).

A central deviation from previous models is the process of *Emergence*, which Mahmoud *et al.* decompose into two individual processes, *Norm Enforcement* and *Norm Adoption*. The latter of those is further decomposed into the processes *Norm Detection* and *Norm Spreading*. Unlike Hollander and Wu's model, emergence is considered a sequential process.

In Mahmoud *et al.* 's model, *Enforcement* consists of direct and indirect sanctioning, where direct sanctioning is the conventional application of reward or punishment, whereas indirect sanctioning is reflected in an individual's reputation and emotions (e.g. guilt).

The *Adoption* process is a composite process that consists of the spreading of new norms and the detection of norms. The *Spreading* process captures the transmission directions outlined by Savarimuthu and Cranefield [2011] (vertical, horizontal and oblique). The *Detection* of new norms captures all forms of norm learning to identify new norms, including imitation, social learning, case-based reasoning and data mining. The model further emphasizes the essential nature of network topologies to facilitate the spreading of norms, including the differentiation of static and dynamic networks, but does not consider alternative mechanisms such as sensing in grid-based environments.

Following the *Emergence* process, the model introduces a novel *Assimilation* process. The authors follow Eguia [2011]'s definition of assimilation "as the process in which agents embrace new social norms, habits, and customs, which is costly but offers greater opportunities" ([Mahmoud *et al.*, 2014b], p.15). In their conception, assimilation involves deciding whether to adopt new social norms by trading off associated costs and benefits.

This process is followed by the *Internalization* process that, similar to Hollander and Wu's conception, includes the *Acceptance*, *Transcription* and *Reinforcement* of the newly acquired norm, with the purpose of embedding it in the agent's behaviour.

The final *Removal* process is equivalent to Hollander and Wu [2011b]'s process of forgetting norms. The purpose is the removal of obsolete norms, as well as being an implicit consequence of norm modification. Mahmoud *et al.* further adopt an unspecified end-to-end process that links *Removal* and *Creation*, possibly implying the evolutionary process introduced by Hollander and Wu.

Discussion The model by Mahmoud *et al.* breaks the trend of proposing progressively more detailed models and attempts to identify the essential processes instead. This condensed conception produces an incoherent understanding of the norm life cycle and semantic ambiguities, the causes of which we will explore in the following section.

Despite the authors' awareness of previous models, in this model emergence only considers the enforcement and adoption of norms (which captures aspects such as spreading and detection), but does not consider the internalization of norms essential for their emergence. How norms can emerge without being internalized is left unexplained. This leaves unclear whether internalization is implied as part of the *Adoption* process that concentrates on spreading and detection of norms. If this were the case, this would produce an ambiguous understanding of the subsequent internalization process.

A similar problem relates to the novel *Assimilation* process, which represents the authors' own substantive contribution [Mahmoud *et al.*, 2014a] to the field of NorMAS. Since assimilation describes the process of deciding whether to adopt given norms, it is unclear in how far this is different from the *Acceptance* process that is part of norm internalization [Mahmoud *et al.*, 2014b], or if it is meant to replace the acceptance component of internalization. The authors' related contribution [Mahmoud *et al.*, 2014a] discusses the assimilation of norms in heterogeneous communities and suggests that the norm internalization itself is a *subprocess* of norm assimilation, an aspect that is not reflected in the sequential organisation of both processes in the life cycle model (see Figure 4). The inspection of the authors' related work suggests that assimilation not so much describes a norm-centred life cycle process. Instead, it characterizes an agent's capability since it describes the *ability and willingness of agents to integrate into their social environment* [Mahmoud *et al.*, 2014a], which entails the adoption of norms, customs, habits, etc.

Overall, the model attempts to rationalize the existing norm life cycle models, leading to a refined but insufficiently specified and contextualized life cycle model, specifically with respect to the emergence process as well as the novel assimilation component – aspects that challenge its coherence and, in consequence, applicability.

2.5 Comprehensive Literature Overview

In the previous sections, we introduced the most relevant life cycle models known in the literature and discussed associated significant contributions. Table 1 integrates the mentioned literature into a comprehensive chronological overview that spans across selected life cycle processes.[6] Whereas the process characteristics of creation, identification, spreading, and enforcement are based on the criteria and approaches discussed in the context of the individual life cycle models (specifically in Sections 2.2 and 2.3), this overview puts particular focus on capturing internalization mechanisms and emergence characteristics, both of which have found limited recognition in previous surveys.

Earlier works on norm internalization apply the specification of norms at design time, which occurs in conjunction with off-line norm creation (which we labelled 'embedded'). However, in the majority of contributions, the adoption and internalization of norms generally occur unreflected (labelled 'immediate'). In more recent approaches, we can observe a shift towards more continuous internalization of norms based on observation ('social learning') as well as probabilistic or threshold-based adoption based on sustained reinforcement ('threshold-based learning', 'Q-learning').

Another category that is characterized by a range of varying, often scenario-dependent measures is the notion of emergence. Examples include convergence thresholds on shared equilibrium strategies in the case of coordination games. In alternative approaches emergence refers to the alignment of sets of norms, both including crisp (e.g. Campenni *et al.* [2009], Andrighetto *et al.* [2010], Griffiths and Luck [2010]) and fuzzy set conceptions (e.g. Frantz *et al.* [2014b; 2016]), or the identification of a shared normative understanding, e.g. by election (Riveret *et al.* [2014]) or by generalization (Frantz *et al.* [2015]). Another group of approaches interpret emergence as the convergence on shared conceptualisations of lexica (e.g. Salazar *et al.* [2010], Franks *et al.* [2013]).

---

[6]This overview refines and extends an earlier survey produced by Savarimuthu and Cranefield [2011].

Publication	Creation	Identification	Spreading	Enforcement	Internalization	Emergence
Axelrod [1986]	-	-	vertical	Sanctions	immediate	Converging strategy choice
Kittock [1995]	-	Machine learning	-	Sanctions	memorizing strategy	Converging strategy choice
Conte and Castelfranchi [1995b]	Off-line design	-	-	-	embedded	Survival under different strategies
Walker and Wooldridge [1995]	-	Machine learning	-	-	-	-
Shoham and Tennenholtz [1992b; 1995]	Off-line design	Machine learning	-	Sanctions	immediate	Converging strategy choice
Shoham and Tennenholtz [1997]	-	-	-	Reputation	embedded	-
Castelfranchi et al. [1998]	Off-line design	-	-	Sanctions	embedded	-
Saam and Harrer [1999]	Off-line design	-	oblique	-	-	-
Verhagen [2001]	Leadership	Machine learning	oblique	Leader/group feedback	alignment with group	Social alignment of action probabilities
Epstein [2001]	-	-	horizontal	-	imitation	Converging on action choice
Flentge et al. [2001]	-	-	vertical	Sanctions	inherited	-
Hales [2002]	Off-line design	-	-	Reputation	immediate	-
Hoffmann [2003]	Entrepreneurship	-	oblique	Reward	immediate	Converging on chosen value
Delgado [2002; 2003]	-	Machine learning	horizontal	Payoff	immediate	Converging on state
López y López and Luck [2004]	Off-line design	Machine learning	-	Sanction/Reward	-	-
Nakamaru and Levin [2004]	-	Machine learning	horizontal	-	immediate	Stabilising norms
Pujol et al. [2005]	-	Machine learning	-	-	-	Converging strategy choice
Chalub et al. [2006]	-	Machine learning	vertical	Reputation	immediate	Converging norms
Fix et al. [2006]	-	-	-	Emotion	immediate	-
López y López et al. [2006; 2007]	Off-line design	-	-	Sanction/Reward	-	-
Sen and Airiau [2007]	-	Machine learning	-	-	immediate	Converging strategy choice
Mukherjee et al. [2007; 2008]	-	Machine learning	-	Payoff	immediate	Converging strategy choice
Savarimuthu et al. [2007; 2008a]	-	Machine learning	-	-	Learning	Converging on value
Campenni et al. [2009; 2010]	-	Cognition, social learning	-	Payoff	-	Shared event-action trees
Savarimuthu et al. [2009]	-	-	oblique	-	immediate	Converging on shared norm
Urbano et al. [2009]	-	Machine learning	oblique	Payoff	immediate	Converging towards joint strategy
Villatoro et al. [2009]	-	-	-	Reward	immediate	Converging action choice
Sen and Sen [2010]	-	Machine learning	horizontal	-	immediate	Converging towards joint action
Griffiths and Luck [2010]	-	-	vertical	-	immediate	Convergence on multiple norms
Savarimuthu et al. [2010b; 2010a; 2011]	Off-line design	Cognition, data mining	-	Sanction signal	immediate	Identification of event sequences as norms
Perreau de Pinninck et al. [2010]	-	-	-	Ostracism	-	Minimal norm violations
Salazar et al. [2010]	-	-	horizontal	-	probabilistic	Convergence on shared word/concept lexicon
Yu et al. [2010]	-	Machine learning	oblique	-	social learning	Convergence on shared goal state
Sugawara [2011]	-	Machine learning	horizontal	Payoff	Q-learning	Convergence on non-conflicting conventions
Villatoro et al. [2011b]	-	-	horizontal	Payoff	social learning	-
Hollander and Wu [2011a]	-	Machine learning	horizontal	internal	threshold-based learning	Equilibrium strategies (various scopes)
Mahmoud et al. [2012; 2015]	-	-	oblique	dynamic sanctions	-	Converging action choice
Riveret et al. [2012; 2013]	-	Machine learning	horizontal	Sanction	Learning	-
Savarimuthu et al. [2013b]	-	Machine learning	-	-	-	-
Villatoro et al. [2013]	-	Machine learning	horizontal	Payoff	Q-learning	Converging action choice
Franks et al. [2013]	-	Evolutionary algorithm	oblique	Leadership	probabilistic	Convergence on shared word/concept lexicon
Morales et al. [2013; 2014; 2015b]	-	Machine learning	oblique	Sanction	immediate	Synthesised set of norms
Mihaylov et al. [2014]	-	Machine learning	horizontal	Payoff	threshold-based learning	Converging action choice
Airiau et al. [2014]	-	Machine learning	horizontal	-	Q-learning	Converging strategy choice
Frantz et al. [2014b; 2016]	-	Machine learning	-	Payoff	Q-learning	Alignment of fuzzy normative understanding
Riveret et al. [2014]	-	Machine learning	oblique	Payoff	Q-learning	Shared normative understanding
Frantz et al. [2014c; 2015]	-	Machine learning	horizontal	Sanction/Reward	Q-learning	Multi-level norm generalization
Yu et al. [2015]	-	Machine learning	oblique	-	Q-learning	Converging strategy choice
Beheshti et al. [2015]	-	Cognition	horizontal	Sanction/Reward	Q-learning	Norm-conforming behaviour

Table 1: Chronological Overview of Literature and Associated Life Cycle Characteristics

2.6 Systematic Comparison of Norm Life Cycle Models

To this stage, we have introduced a diverse set of life cycle models along with associated contributions, but have yet to relate those systematically. Finnemore and Sikkink [1998] 's model (Section 2.1), proposed in the field of international relations, identifies three processes in a norm's life, starting with its explicit creation (*Emergence*), its spreading (*Cascade*), leading to wide-ranging adoption (*Internalization*). In contrast to all other models, their model looks at states as central players and emphasizes the long-term perspective of normative change (e.g. embedding the changing societal normative view in professional ethics).

The remaining three ones are products of systematic reviews of contemporary research in the area of NorMAS, an approach spearheaded by Savarimuthu and Cranefield [2011] . Their model (Section 2.2) provides a refined account of the beginning of a norm's development, with a particular focus on the initial formation and propagation. Their model interprets emergence as an outcome measure and does not include a long-term perspective on norms, such as their decay and substitution over time.[7] However, since their model is grounded in a systematic review of existing works, this does not indicate a principle shortcoming of the model, but rather reflects the contemporary state of the research field.

Hollander and Wu [2011b]'s model (Section 2.3) provides the most comprehensive account of norms' life cycles, and, similar to Savarimuthu and Cranefield's grouping of processes into stages, identifies essential superprocesses that are composed of refined subprocesses. Their model goes beyond previous accounts and proposes processes that are only weakly reflected in literature, thus identifying presumed research gaps. The most important contribution of their model is the recognition of cycles of recurring processes, an example of which is the characterisation of norm emergence as a reiteration of transmission, enforcement and internalization. The second essential contribution is the integration of a long-term perspective on normative change, which they reflect as an evolution process.

Finally, Mahmoud *et al.* [2014b] (Section 2.4) describe a model that condenses the number of relevant processes of the normative life cycle to five. Their model puts specific emphasis on norm assimilation, i.e. an individual's decision whether to accept (and subsequently internalize) a given norm. They further decompose the emergence process into enforcement and adoption (which in itself consists of the processes *Norm Spreading* and *Norm Detection*). Similar to Finnemore and Sikkink, as well as Savarimuthu and Cranefield, Mahmoud *et al.* conceive a linear norm life cycle; they do not consider iterative processes.

An aspect that challenges the systematic comparison of all four models is not only the varying level of detail, but the observable terminological ambiguity. In the different life cycle models the sharing or spreading of norms is selectively captured by the terms 'cascade' (Finnemore and Sikkink), 'spreading' (Savarimuthu and Cranefield, Mahmoud *et al.*), and

---

[7]They consider those as part of a refined set of life stages in later work [Savarimuthu *et al.*, 2013b].

'transmission' (Hollander and Wu). A further notable example is the norm 'identification' (Savarimuthu and Cranefield) that is alternatively characterized as 'recognition' (Hollander and Wu) or 'detection' (Mahmoud *et al.*).

Beyond those synonyms, the specific processes in different models have semantic overlappings. To facilitate a systematic comparison of content and semantic relationship, in Figure 5 we provide an overview of all life cycle models, with individual processes roughly aligned by semantic relationship. Process labels are formatted and grouped to reflected their nature and importance in the respective life cycle model:

- Savarimuthu and Cranefield differentiate between individual processes and stages. Consequently, the *life cycle stage names* are held in bold font.

- Hollander and Wu's *superprocess labels* are held in bold font. The emergence and evolution processes are further explicitly included in the schematic overview.

- Mahmoud *et al.*'s model composes the emergence process from two elementary processes and is thus held in bold font, along with all further *processes of the same conceptual weight*.

Dotted lines indicate the semantic relationships between individual processes of the corresponding life cycle models. For example, Finnemore and Sikkink's *Cascade* process combines components of Savarimuthu and Cranefield's *Spreading* and *Enforcement* processes.

Despite the diversity of norm life cycles, the systematic review of all models reveals clusters of processes that have similar or identical functions (identified as solid horizontal lines in Figure 5). We can generalize four such clusters, or phases, of norm life cycles, and label those by complementing the labels of the initial two life cycle stages in Savarimuthu and Cranefield [2011]'s model:

- Formation – Processes associated with the creation and inference of norms

- Propagation – Processes associated with the communication of norms

- Manifestation – Processes associated with the general acceptance and entrenchment of norms

- Evolution – Processes associated with the evolutionary refinement of norms

The identified phases correspond to the abstract phases proposed by Andrighetto *et al.* [2013], namely *Generation*, *Spreading*, *Stability* and *Evolution*, an aspect that supports the semantic process clusters proposed above. Our terminological choice is driven by the goal

518

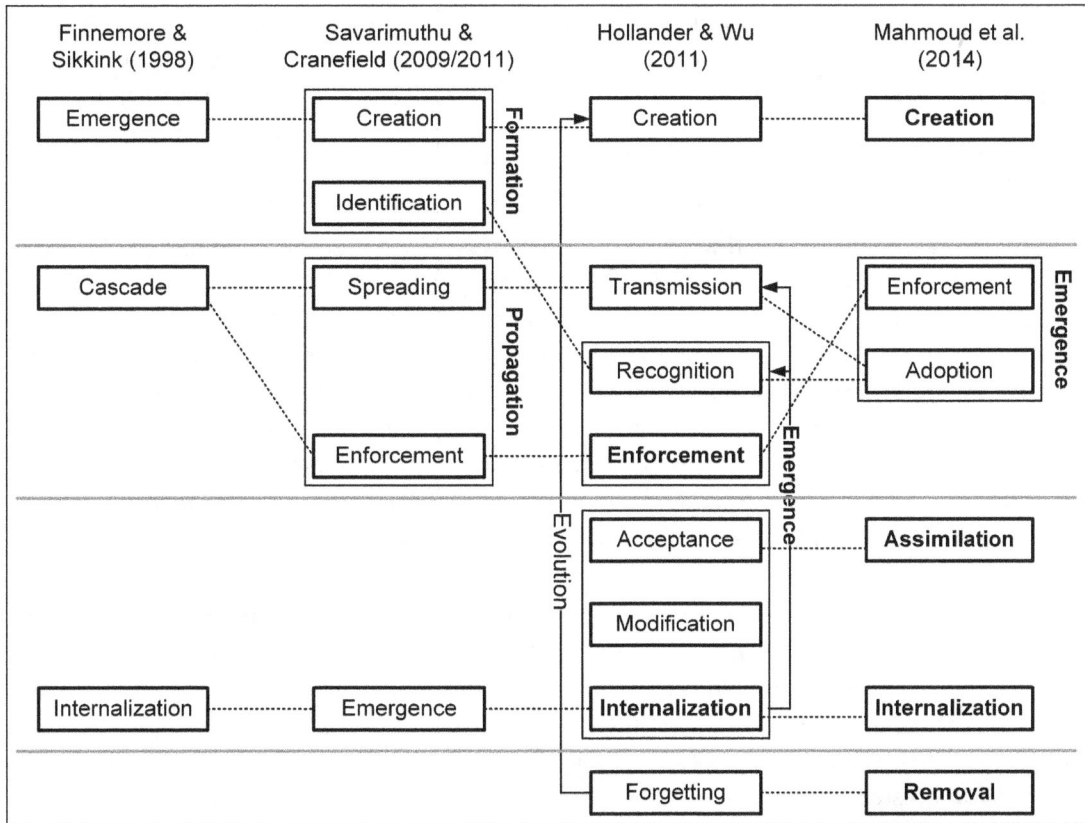

Figure 5: Schematic Comparison of Discussed Norm Life Cycle Models

to comprehensively capture the semantics of associated processes of all discussed norm life cycle models (e.g. operations associated with norm internalization extend beyond the characterization of a norm as stable – see discussion below). In the following, we will use the identified phases to compare and contextualize the norm life cycle models.

Phase 1: Formation All models identify norm creation as the initial life cycle step. In contrast to all other models, Finnemore and Sikkink [1998] employ a different emergence understanding. In their conception emergence entails the initial creation of a norm (which Hollander and Wu [2011b] describe as "norm creation on a micro scale" [Hollander and Wu, 2011b]), whereas life cycle models from the area of NorMAS (henceforth referred to as NorMAS models) understand emergence as "norm establishment on a macro scale" [Hollander and Wu, 2011b]. However, the underlying understanding of this initial phase – the explicit creation of a norm – is identical for all models. Despite this uniform characterization, we label this phase as *Formation* in order to capture a more general understanding of norm creation, widening the scope to approaches that do not rely on explicit norm creation such as the identification of existing/unknown norms by observation, an aspect implicitly captured by Savarimuthu and Cranefield's notion of *Norm Identification* (which we discuss in Section 4).

Phase 2: Propagation Following the creation, all models describe some sort of norm communication, or propagation (*Cascade*, *Spreading*, *Transmission*, and *Adoption*). A special case is Mahmoud *et al.* [2014b]'s *Adoption* process, which entails both norm spreading and detection. All NorMAS models recognize a notion of norm identification (*Identification*, *Recognition*, and *Adoption*), but have a varying sequential organisation. While Savarimuthu and Cranefield's early allocation of norm identification is driven by the understanding that agents need to identify norms in their environment, all subsequent models interpret it as a step that follows the transmission of a norm. Similarly, all NorMAS models recognize enforcement as an essential determinant of a norm's success.

Phase 3: Manifestation The propagation of norms is followed by their *Internalization*. In Finnemore and Sikkink's model that refers to the wide-ranging adoption of a norm within society and its embedding in societal institutional structures. In addition to gaining stability, at this stage norms thus manifest themselves in the social fabric which implicitly reinforces their persistence, constrains future action, but also limits the potential of competing norms. Manifested norms can attain quasi-legal status, e.g. by shaping the codes of ethics for specific occupations, which are subsequently absorbed into the discipline's professional training and practices. This understanding is compatible with Savarimuthu and Cranefield's *Emergence* interpretation, which represents the extent to which a norm is able to penetrate the affected society.

While these first two models describe norm manifestation as a macro-level process, the models of Hollander and Wu as well as Mahmoud *et al.* describe refined sets of micro-level processes that lead to the internalization of norms. Hollander and Wu differentiate between *Acceptance*, *Modification*, and *Internalization*, including the decision whether to adopt a norm in the first place, and reflecting individual biases introduced during internalization. Mahmoud *et al.* reduce those to two processes, namely *Assimilation* and *Internalization*. As discussed in Section 2.4, the authors borrow the notion of *Acceptance* (which is identical to Hollander and Wu's *Acceptance*[8]), and consider it part of the *Internalization* process.

---

[8]"Norm acceptance is a conflict resolution process in which external social enforcements compete against the internal desires and motivations of the agent. If the new norm is in conflict with existing norms and may lead to inconsistent behaviours, or if the cost of accepting the new norm is too high, it will be rejected ..." [Hollander and Wu, 2011b], paragraph 3.24.

However, they introduce a preceding *Assimilation* process[9] (whose function is not clear, since it is not sufficiently contrasted to *Acceptance*) and *Internalization*. At the end of this manifestation phase, all models assume that individuals have embraced the promoted norms.

Phase 4: Evolution   The fourth phase which we tag *Evolutionary Phase* is only reflected in the later life cycle models which introduce the processes *Forgetting* and *Removal* that reflect the end of the normative life cycle. However, more important than their function to 'complete' the norm life cycle is their role as starting point for an evolutionary process (as introduced by Hollander and Wu [2011b]; Section 2.3) in which norms are refined or substituted by more relevant or efficient norms; forgetting old norms is a by-product of this evolutionary refinement and technical necessity to maintain efficient but also realistic architecture implementations.

The 'Special Case' Emergence   Only exception to the uniform organisation of processes into general phases is the notion of emergence, which reflects the terminological ambiguity surrounding this concept. Whereas Finnemore and Sikkink's micro-level interpretation of emergence is associated with the *Formation Phase*, Mahmoud *et al.* see the *Propagation Phase* with the processes of *Enforcement* and *Adoption* as decisive for emergence. Hollander and Wu see emergence as an iterative process that spans across *Formation* and *Manifestation Phase*. Savarimuthu and Cranefield associate emergence with the third phase of norm manifestation and interpret it as a result of *Formation* and *Propagation*.

We believe that Hollander and Wu's cyclic representation represents the most accurate characterisation of the emergence process, since it links the macro-level emergence process with the underlying propagation and internalization processes, an aspect we will revisit in the context of proposed refinements (see Section 2.7). Savarimuthu and Cranefield's interpretation as outcome measure only reflects a quantifiable macro-level phenomenon, but does not maintain its relationship to the underlying processes that produce it. Mahmoud *et al.* inherently rely on propagation processes to determine a norm's emergence. Their model neither considers the cyclic nature of emergence nor does it consider the internalization of norms as a precursor for their further spread (see discussion in Section 2.4).

Norm Life Cycle Models and Levels of Analysis   Comparing the individual models leaves the general impression that later models (with exception of Mahmoud *et al.*) are increasingly detailed and comprehensive. However, while this observation is warranted, it rather reflects the operational levels the life cycle models represent. Finnemore and Sikkink's,

---

[9]"[Norm assimilation is] ... the process in which agents embrace new social norms, habits and customs, which is costly but offers greater opportunities." [Mahmoud *et al.*, 2014b], p.15 with reference to Eguia [2011].

as well as Savarimuthu and Cranefield's models, describe the adoption and implementation of norms on the macro level, i.e. group or society level. This is well captured in Finnemore and Sikkink's understanding of internalization as the process of embedding the norm in a society's social structures and institutions. Similarly, Savarimuthu and Cranefield describe emergence as a macro-level outcome that describes the adoption of a norm across the wider society. Hollander and Wu's model introduces a shift from the macro-level norm perspective to an individual-centred micro-perspective, an aspect that is particularly apparent in the elementary processes they describe in the context of the establishment phase. Micro-level processes include *Acceptance* (the decision whether or not to accept norms), *Modification* (the modification of norms during internalization based on individual biases), and finally *Internalization*, which describes an individual's integration of norms into its existing belief structure. Only the subsequent *Emergence* and *Evolution* processes operate on the macro level, since they shift the perspective from individual to society level. Mahmoud *et al.*'s model similarly emphasizes individual-level processes such as *Assimilation* and *Internalization*, which they decompose into operational steps that are similar to Hollander and Wu's processes (Mahmoud *et al.*: *Acceptance, Transcription, Reinforcement*; Hollander and Wu: *Acceptance, Modification, Internalization*). In both models forgetting and removal of norms emphasizes a micro-level operation and is considered a technological necessity (in the light of limited computational resources), but obscures the macro-level function of facilitating an evolutionary refinement [Hollander and Wu, 2011b] of the normative landscape.

Understanding the different operation levels of the introduced models is helpful, since it allows their selective consultation. For the modelling and analysis of macro-level phenomena, the use of Savarimuthu and Cranefield's model may provide sufficient conceptual backdrop, whereas detailed cognitive agent models will find the most comprehensive structural blueprint in Hollander and Wu's model, with other models providing even higher levels of abstraction (Finnemore and Sikkink) or varying emphasis of individual-level processes (Mahmoud *et al.*).

2.7   General Norm Life Cycle Model

As a result of reviewing the existing life cycle models and their respective biases, we propose a general life cycle model that harmonizes various inconsistencies of the introduced approaches (e.g. micro- vs. macro-level operation, emergence understanding), but also addresses explicit conceptual omissions that are of increasing importance in recent developments (see Sections 3 and 4).

As such, the proposed general norm life cycle model introduces five essential revisions, which we discuss in the following:

- Distinction between micro-level processes and macro-level phenomena

- Norm Identification as an alternative life cycle entry point (in addition to explicit norm creation)

- Enforcement as a dynamic process with norm emergence as a resulting phenomenon

- Norm Forgetting as by-product of norm evolution

- Potential norm modification throughout all life cycle processes

Distinction between Micro-Level Processes and Macro-Level Phenomena   As discussed█ in great detail in the previous Section 2.6, the existing norm life cycle models operate on varying levels of abstraction, with the initial models identifying coarsely-structured processes, whereas the latter two models describe processes of varying granularity (e.g. Hollander and Wu's end-to-end processes, superprocesses in addition to regular processes). We propose a systematic distinction by separating the micro-level processes (e.g. Transmission, Identification and Internalization) that find explicit representation in normative architectures, from macro-level phenomena that arise from the cyclic operation of the underlying processes. We believe that differentiating between a processual and phenomenological perspective on norms is useful to inform modelling considerations in different problem domains, such as the engineering of a process-centric normative agent architecture, in contrast to macro-level processes such as the emergence of norms within agent societies or their evolution over time. However, at the same time, these perspectives should not be dissociated in order to retain the links between the phenomena and the underlying processes. *Norm Emergence* is thus a result of iterative Transmission, Identification, Internalization and Enforcement processes. *Norm Evolution* extends across the entire norm life cycle, additionally involving the inception of new norms (Norm Creation) as well as the forgetting of decaying norms (Norm Forgetting).

Norm Identification as a Life Cycle Entry Point   To date, the existing approaches assume the explicit creation of a norm. Proposed mechanisms include norm leadership, entrepreneurship, autonomous innovation and social power. However, in reality, norms may not necessarily be explicitly created, of unknown origin, but be rooted from behavioural regularities based on individuals' necessity to act in the first place (described as "urgency of practice" [Bourdieu, 1977]). In principle, a situational strategy choice to coordinate behaviour (e.g. chosen means of greeting, road-side choice) can emerge as self-enforcing convention (without intentional explicit conceptualisation), before finding recognition as a fully fledged norm.[10] Previous works acknowledge the existence of natural emergence pro-

---

[10]Examples for works that showcase this characteristic (e.g. Morales *et al.* [2015a], Riveret *et al.* [2014], Frantz *et al.* [2015]) are discussed in the context of the upcoming Section 4.

cesses[11] (Boella *et al.* [2008], Finnemore and Sikkink [1998], López y López *et al.* [2007], Savarimuthu and Cranefield [2009]), but assume an explicit creation as the starting point of the normative life cycle. We propose that a comprehensive norm life cycle should reflect the unplanned inception of norms based on social interaction as a possible alternative starting point of a norm's life – in addition to the explicit creation.

Enforcement as a Dynamic Process   A further aspect relates to the role of enforcement. All NorMAS life cycle models represent enforcement as an explicit process that appears independent of notions such as spreading. However, enforcement itself can be interpreted as a dynamic process that promotes the cyclic reinforcement of norms, leading to their spread and thus their increasing adoption, producing emergence as an associated phenomenon (as discussed in the previous paragraph). Some form of enforcement – whether implicitly (e.g. serving as a guiding role model or influence based on shared values) or explicitly (e.g. by engaging in overt sanctioning) – is a prerequisite for the transmission of norms. In this context, it is further important to note that enforcement does not carry a specific valence, but can bear positive associations, such as providing a reward for a norm-compliant employee, or represent an explicit punishment, such as humiliating an individual in front of her reference group (e.g. an employee amongst fellow co-workers). Apart from such forms of overt *external enforcement*, enforcement can further be directed at oneself (internal enforcement), reflected in emotions such as the "warm glow" [Andreoni, 1989] of compliance (i.e. 'doing the right thing') or the guilt of violation (e.g. engaging in jaywalking despite conventional compliance).

Whether implicit or explicit, positive or negative, internal or external, enforcement relies on the prior internalization by the potential enforcer. This does not necessarily imply that the enforcer applies this norm to her- or himself or even 'believes' in it. As such, individuals can be tasked with the enforcement or feel pressured to defend norms they object to (such as not engaging in jaywalking in the presence of bystanders). Similarly, not violating a norm when facing the opportunity (without actively promoting it) can act as norm reinforcement. An example for this is the rejection of a bribe, especially if the actor holds a role model function (e.g. as a manager) [Hogg, 2001]. Conversely, the observation of violation by an authority figure (e.g. taking a bribe) can accelerate norm erosion. Whether compliant or not, essential for any positive or negative enforcement is some internalized conceptualization of the enforced norm in order to make its compliance and violation detectable. Consequently, we do not see emergence as a process in itself, but as a phenomenon that results from a sustained cyclic reinforcement based on the transmission, identification, internalization, and subsequent enforcement of norms, leading to their manifestation.

---

[11] Here emergence should be understood as the micro-level process of norm inception.

Forgetting as a By-Product of Norm Evolution   A final aspect relates to the notion of forgetting. Hollander and Wu introduce forgetting as an end point of an evolutionary cycle that affords a norms refinement. However, the conceptualisation as an 'end-to-end process' presents it as a sequential step in a series of processes. Similar to the conception of emergence laid out before, we see evolution as a phenomenon that arises from the continuous reinforcement of norms, their change during identification and internalization, as well as their potential to become obsolete and ultimately forgotten. This process cannot be conceived as sequential but operates concurrently, with newly identified norms gaining more salience and potentially leading to existing norms' adaptation or decay. Though forgetting is an essential endpoint in the normative life cycle, it does not represent the starting point for a continuously operating evolution process; 'forgetting' is a by-product of evolving norms.

Schematic Overview   In Figure 6 we show a schematic overview of the proposed refined norm life cycle that condenses elements of the previously introduced models, but incorporates essential revisions. We will briefly explore the processes in the following.

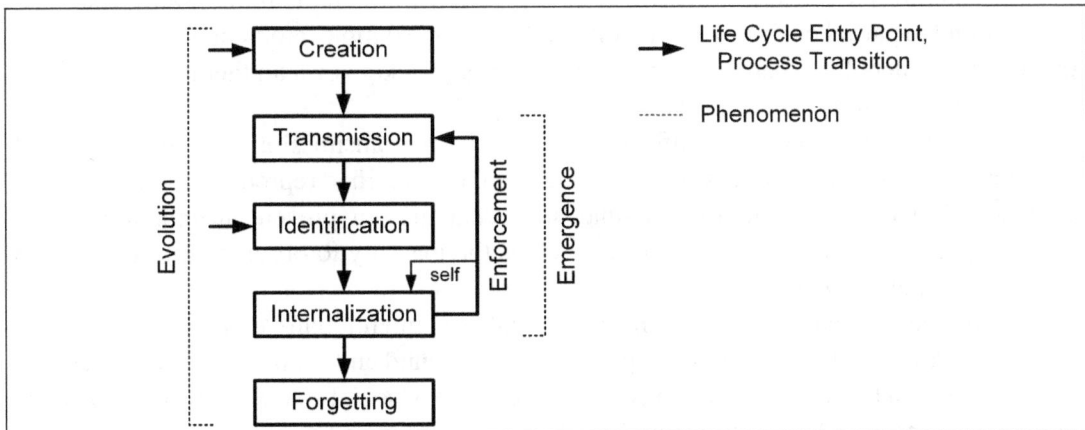

Figure 6: General Norm Life Cycle

As stated before, norms can either be explicitly created or identified at runtime (the corresponding right-facing arrows in Figure 6 mark these life cycles starting points). If created, norms are transmitted and identified.[12] As mentioned above, identification is not only initiated by transmission, but may involve the identification of an existing norm (e.g. by observation). Once internalized (by a complex internalization process that may contain elementary processes as laid out by Hollander and Wu [2011b]), norms can be reinforced, which may operate internally (e.g. based on motivational enforcement or elicited emotions),

---

[12]Note that we use terms synonymously for the ambiguous terminology in existing life cycle models as discussed before. In this case, the notion of 'identification' is identical to 'recognition'.

or be directed towards external targets. External enforcement requires the transmission of normative content, the subsequent identification and internalization by enforcement targets, and so on. This constitutes the norm's emergence. At any time, new norms can be created or identified, potentially causing change in the normative system by emerging and becoming salient. If cyclic reinforcements of a given norm cease, the norm loses its relevance and is incrementally forgotten. This second phenomenon can be understood as norm evolution. Both, emergence and evolution, are similar in that they represent phenomena (and could be construed as meta-processes in the epistemological sense[13]), but they vary in scope regarding the involved processes.

Norm Modification throughout Norm Life Cycle Hollander and Wu [2011b] discuss the modification of norms as part of the internalization process. However, we believe that the potential for norm modification, whether intentionally and systematic or not, arises during *any* form of transmission, internalization, or subsequent externalization (e.g. enforcement) of normative content.

This can involve the loss of information during transmission or simply transmission errors, leading to partial or simply wrong information. For example, ambient traffic noise may prevent bystanders from perceiving the scolding of jaywalkers or lead them to misconstrue the normative content (e.g. as a heated discussion).

Complementing potential modification sources during transmission, the identification of norms can be challenged by sensory biases that lead to a modified reproduction of normative content. Visual impairment, for example, may challenge or prevent an individual from capturing normative signals of relevance, such as the inability to observe a norm violation in the form of jaywalking.

During the internalization of norms, individuals can intentionally modify their interpretation of norms based on individual experience, background and aspirations. Hand-shaking, for example, can be interpreted as an acknowledgement of social status or objected to on the grounds of potential disease transmission. While the perceived action may be unambiguous (i.e. not manipulated during transmission and sensing), the individual may introduce an intentional bias, such as building a negative connotation with an internalized norm with an intent to change or abandon it.

This subjective perception of social reality extends to the unconscious realm, with an abundance of further mechanisms at work that drive individuals' biases in decision-making, belief formation and behaviour, as well as memory and social biases. Decision-making biases can be introduced by the oftentimes disproportionate perception of rewards and sanc-

---

[13]Our interpretation is in contrast to Hollander and Wu (Section 2.3) who use the term to describe end-to-end connections between elementary processes. They essentially consider regular processes and end-to-end processes as same natural kinds, and consequently do not allocate the operation of meta processes on a higher level of abstraction.

tions as well as an asymmetric risk tolerance (see e.g. Prospect Theory [Kahneman and Tversky, 1972]). An illustrative fact in line with this observation is that individuals are by magnitudes of thousands more likely to succumb to diseases from behavioural causes (e.g. lack of exercise, smoking) than terrorist attacks, yet fear the latter disproportionally more.

Further behavioural biases, for example, include paying selective attention to favourable information, as well as seeking for confirmation of conceptions and beliefs that we already hold (confirmation bias), such as the focus on information that 'validates' an existing norm. Memory biases are fundamentally concerned with humans' limited information processing capabilities (bounded rationality [Simon, 1955]), including limited information recall, the fading of memory over time, as well as our brain's ability to fill in of memory from imagination (false memories), all of which can lead to the distortion of internalized norms. Similarly, social effects can lead to biases with respect to the normative content, such as biases towards conformity with authority figures or ingroup members. Many of these systemic biases interact with human mechanisms for operating under uncertainty. Examples for such mechanisms include the use of stereotypes to ascribe characteristics to unknown individuals (implicit social cognition [Greenwald *et al.*, 2002]), or the application of irrational decision-making heuristics when acting under pressure ('gut feeling').

The presented selection of the cognitive biases is non-exhaustive, of course, but it offers a starting point for the exploration of cognitive influences that distort the interpretation of normative content during the norm internalization process.

Finally, norms can be modified based on the *characteristics of enforcement and enforcer*, generally affecting the salience and predictability of norms.

One fundamental determinant is the *valence* of enforcement, i.e. whether a norm is reinforced by rewards (such as a 'pat on the back') or punishments (such as scolding). As indicated earlier in the context of discussing cognitive biases, the nature of enforcement can modify norms. This includes the asymmetric impact of positive and negative sanctions (see e.g. Kahneman and Tversky [1972] and Baldwin [1971]), but also frequency, intensity and variation in enforcement. Infrequently reinforced norms are unlikely to gain high salience and may thus be easily foregone. Highly variable or inconsistent enforcement, however, interacts with individuals' risk affinity (e.g. promoting probabilistic norm compliance) but also involves the perceived level of fairness (e.g. inconsistent leadership behaviour in organisational environments [Sims and Brinkmann, 2003]), which can lead to the loss of norm commitment by norm subjects, or even active opposition.

Other influence factors on enforcement that can lead to norm modification include the *social relationship between enforcer and subjects*, but also the *nature of the enforcer*. As shown by Goette *et al.* [2006] and Horne [2007], increased social relationship (e.g. shared group membership) between enforcer and subjects correlates with the enforcement practice. However, the central or distributed nature of the enforcer can be decisive for the enforcement. Enforcers can be quasi-centralized and self-appointed (e.g. such as rules regarding dish washing procedures imposed by administrative secretary) and show predictable enforcement strategies ('conventional sanctions'), whereas decentralized enforcement can be unpredictable with respect to the number of enforcers (e.g. unknown number of enforcers objecting to jaywalking), the applied strategies (e.g. gestures vs. scolding) and emerging dynamics (e.g. eruption into collective participation in humiliation), and thus lead to nuanced reinforcement and conceptualisation of the norm as more or less serious.

Complementing the misinterpretation of normative content based on sensory bias, enforcers can likewise cause a modification of normative content by sending ambiguous signals. Examples include the insufficient command of language to express a sanction appropriately or the confusion of terminology for reward and punishment (e.g. 'awesome' vs. 'awful').

Table 2 highlights the discussed potential causes for norm modification and associates those with individual processes. While this selection identifies potential modification sources, specific factors depend on the scenario, the capabilities of the transmission medium, as well as sensory and cognitive agent models and corresponding action capabilities. In addition to intentional modification, norms can thus essentially be modified whenever an individual interacts with its social environment, the effects of which can accumulate and drive the continuous evolution norms are subjected to, providing a starting point for exploring the emergence of divergent norms within separated social clusters.

Process	Causes for Modification
Transmission	Information Loss; Transmission Errors
Identification	Sensory Biases/Constraints
Internalization	Cognitive Biases; Intentional Modification
Enforcement	Choice of Enforcement; Characteristics of Enforcer(s); Relationship to Enforcement Target

Table 2: Potential Sources of Norm Modification

Summary In this section, we have proposed a general norm life cycle model that builds on the systematic comparison of existing life cycle models, harmonizes identified terminological and conceptual inconsistencies (see Section 2.6 for details), and introduces additional characteristics we deem relevant for a *general* norm life cycle model (e.g. norm identification as an alternative life cycle entry point).

While this proposed model highlights the essential processes of a general norm life cycle that we believe are necessary for its operationalization, it leaves the potential for the domain- or model-dependent refinement of individual processes, similar to Hollander and Wu's model. However, this model integrates the commonalities of existing models, while offering a comprehensive and consistent reflection of norm dynamics found in the contemporary literature. It further provides a clear differentiation between processes and associated phenomena.

2.8 Discussion

Based on the condensed, yet comprehensive overview of selected existing normative life cycle models[14], we provided a systematic comparison and synthesized the identified essential components into a refined interpretation of the normative life cycle. However, the focus on individual processes of the life cycle models obscures two areas of development that *combine* individual processes to model norm dynamics comprehensively – the areas of *norm change* and *norm synthesis*. We will explore those specific areas in the following, before contextualising those with the proposed life cycle concept at the end of this article.

3 Norm Change

3.1 Overview

In the previous sections, we have seen different models that have been introduced in the literature to capture the life cycle of norms. These models consider the creation of norms, the processes that can facilitate their spreading, and the recognition (or learning) of norms by agents. Yet, we also know that in human societies norms can *change* over time. For example, on the occasion of the G8 summit in 2009 in Italy the Schengen treaty was suspended to guarantee the security of the local population and of the delegations, and then reinstated. In a similar way, normative systems in multi-agent systems must be able to evolve over time, for example due to actions of creating or removing norms in the system. However, the dynamic nature of norms in artificial systems is often not addressed in the simulation work on norms.

Norms are crucial in modeling agents' interactions. The definition of a normative multi-agent system that the community put forward at the first NorMAS workshop in 2005 is that "Normative MultiAgent Systems are multi-agent systems with normative systems in which agents can decide whether to follow the explicitly represented norms, and the normative systems specify how and in which extent the agents can modify the norms" [Boella *et al.*, 2006]. In order to ensure systems with autonomous agents, it is essential that norms can be

---

[14]Further life cycle models include the ones proposed by Andrighetto *et al.* [2013] and Singh [2014], but have been excluded from this comparison because of their highly abstract perspective or fine-grained computational focus.

violated (even though non-compliant agents are sanctioned). Because of the accent on the ability of the agents to modify norms, this definition was then known as "the normchange definition" of normative multi-agent systems.

The central problem of changing norms lead to two workshops on the dynamics of norms, the first one in 2007 in Luxembourg and the second one in 2010 in Amsterdam[15]. These two international workshops brought together researchers working on norm change from different perspectives. The revision of norms was also one of the ten open philosophical problems in deontic logic highlighted in Hansen *et al.* [2007] and further extended in Pigozzi and van der Torre [2017]. As we will see in the pages that follow, a consensus on a common framework to model norm change is still lacking.

3.2   From Law to Logic

Historically, the first approaches to norm change were driven by lawyers. For instance, at the 1981 international conference 'Logica, Informatica, Diritto' held in Florence (Italy), one of the conference sessions was explicitly dedicated to the problem of the abrogation of rules[16]:

> The abrogation of rules creates special problems in determining which is the 'legal system in force', as in the case of abrogation of the consequences of explicit rules and not of the rules themselves.

In the same years, a logic study of the changes of a legal code brought together three researchers coming from different backgrounds: Alchourrón, Gärdenfors and Makinson, respectively a legal theorist, a philosopher and a logician.

At the beginning, it was Alchourrón and Makinson who started investigating three types of change (Alchourrón and Makinson [1981; 1982]). The first type consists of the addition of a new norm (consistent with the other norms in the code) to an existing code. Such enlargement leads to the addition of the new norm to the code along with all the consequences that can be derived from it. The second type of change occurs again when a new norm is added, but now the new item is inconsistent with the ones already in the code. In this case we have an *amendment* of the code: in order to coherently add the new regulation, we need to reject those norms that conflict with the new one. Finally, the third change occurs when a norm is eliminated (technically, a *derogation*). In order for the elimination to be successful, however, also all other norms of the existing code that imply that norm have to be eliminated.

---

[15]http://www.cs.uu.nl/events/normchange2/

[16]When a norm is abrogated, its effects in the past still hold. This is different from the annulment of a norm, which also eliminates its effects in the past.

The approach of Alchourrón and Makinson was general: in the definition of change operators for a set of norms of some legal systems, the only assumption was that a norm is a formula in propositional logic. Thus, they suggested that "the same concepts and techniques may be taken up in other areas, wherever problems akin to inconsistency and derogation arise" ([Alchourrón and Makinson, 1981], p.147).

When Gärdenfors joined (at that time he was mainly working on counterfactuals), the trio became the founders of the well-known AGM theory, and started the fruitful research area of belief revision [Alchourrón et al., 1985]. Belief revision is the formal study of how a theory (a deductively closed set of propositional formulas) may change in view of new information, which may cause an inconsistency with the existing beliefs.

Expansion, revision and contraction are the three belief change operations that Alchourrón, Gärdenfors and Makinson identified. *Expansion* is the addition of a new proposition that is not in conflict with the existing formulas in the theory. *Revision* is the addition of information that is inconsistent with the existing beliefs. In order to consistently add such information, all conflicting formulas have to be removed. Finally, *contraction* is the elimination of a formula from the theory.

The AGM theory provides a set of postulates for each type of theory change. There is an obvious correspondence between the three types of belief change and the three changes in a system of norms mentioned above. The link between theory change and change of a legal code was explicitly acknowledged by Alchourrón, Gärdenfors and Makinson:

> [...] theory *contraction*, where a proposition x which was earlier in a theory A, is rejected. When A is a code of norms, this process is known among legal theorists as the *derogation* of x from A. [...] Another kind of change is *revision*. [...] In normative contexts this kind of change is also known as *amendment*. ([Alchourrón et al., 1985], p. 510)

It should be noted, however, that the AGM theory was mainly used for belief change. This is because beliefs and norms were both represented as formulas in propositional logic.

One of the first attempts to specify the AGM framework to tackle norm change was a paper by Maranhão [2001], presented at the 2001 ICAIL conference. The approach was inspired by Fermé and Hansson [1999]'s selective revision, where only part of the input information is accepted. Maranhão introduced a *refinement* operator, which refines an agent's belief set by accepting the new input under certain conditions. Refinement provides a tool to represent the introduction of exceptions to rules in order to avoid conflicts in normative systems (for instance in those cases where judges face new conditions which were not mentioned in the legal statute but turn out to be relevant in practical situations).

As we will see in the following pages, the belief revision approach has been recently reconsidered to represent and reason about norm change (see Section 3.4).

3.3 Semantic Approaches

Two main approaches to model norm change have been developed in the literature: semantic approaches inspired by the dynamic logic approach [van Ditmarsch and van der Hoek, 2007], and syntactic approaches where norm change is performed directly on the set of norms.

Among semantic approaches we find the dynamic context logic proposed by Aucher *et al.* [2009], which represents norm change (in particular the dynamics of constitutive norms[17]) as a form of model update. Starting from a modal logic of context [Grossi *et al.*, 2008], context expansion and context contraction operators are introduced. The intuition is that contexts can be seen as set of models of theories. Context expansion is thus linked to the promulgation of counts-as conditionals while context contraction is used for the abrogation of constitutive norms. Norms are statements of the kind "the fact α implies a violation". One of the advantages of this approach is that it can be used for the formal specification and verification of computational models of interactions based on norms.

A similar proposal is by Pucella and Weissman [2004], where operations for granting or revoking extensions are defined in a dynamic logic of permission. Aucher *et al.* [2009]'s framework is more general. Changes in the granting and revoking of permissions and obligations are more specific than the normative system change captured in Pucella and Weissman [2004]'s article.

3.4 Syntactic Approaches

3.4.1 Defeasible Logic

When new norms are created or old norms are retracted from a normative system, the changes have repercussions on obligations and permissions that such norms established. Obligations can change without removing or adding norms. For example, change in the world can lead to new obligations without changing the legal norms. For this reason, Governatori and Rotolo [2010] insist on the need to distinguish norms from obligations and permissions (as done in deontic logic).

Inspired by the legal practice, Governatori and Rotolo aim at a formal account of legal modifications. They use a syntactic approach, where norm change is an operation performed on the rules contained in the code. Such modifications can be implicit or explicit. Implicit modifications are the most common. They arise when new norms are introduced in the legal system and such norms conflict with existing ones. The new norms enforce a retraction of the old ones because, for example, have a higher ranking status, like a national law can

---

[17]Constitutive norms are rules that define an activity. For example, the institutions of marriage, money, and promising are systems of constitutive rules or conventions. As another example, a signature may count as a legal contract, and a legal contract may define a permission to use a resource and an obligation to pay.

derogate a regional law. Explicit modifications are obtained when norms that define how other existing norms have to be modified are added to the legal code.

In particular, the mechanisms of annulments and abrogations are studied. Annulment removes a norm from the code. It operates *ex tunc*: all effects (past and future) are cancelled. Abrogation too is a kind of norm removal but, unlike annulments, it applies *ex nunc*: it cannot operate retroactively, leaving their effects in the past hold.

The notion of abrogation is complex and there is no agreement among jurists on whether abrogations actually remove norms or not. In order to illustrate the difficulties, Governatori and Rotolo give the following example:

> If a norm n_1 is abrogated in 2007, its effects are no longer obtained after then. But, if a case should be decided in 2008 but the facts of the case are dated 2006, n_1, if applicable, will anyway produce its effects because the facts held in 2006, when n_1 was still in force (and abrogations are not retroactive). Accordingly, n_1 is still in the legal system, even though is no longer in force after 2007. ([Governatori and Rotolo, 2010], p. 159)

As seen in this example, the difficulty of abrogations comes from the fact that, in most cases, direct effects should be removed, but this is not necessarily the case for indirect effects. Clearly the temporal dimension is crucial in their formal representation, but it also makes the formalisation more cumbersome.

So Governatori and Rotolo first try to capture annulments and abrogations with theory revision in defeasible logic without temporal reasoning. Unfortunately, the result is not fully satisfactory as retroactivity cannot be captured. This is a crucial aspect as retroactivity allows to distinguish abrogation from annulment.

In the second part of the paper then, they use a temporal extension of defeasible logic to keep track of the changes in a normative system and to deal with retroactivity.

Norms have two temporal dimensions: the time of validity of a norm (when the norm enters in the normative system) and the time of effectiveness (when the norm can produce legal effects). As a consequence, multiple versions of a normative system are needed. In order to illustrate the problem, we recall this example from a hypothetical taxation law discussed in [Governatori and Rotolo, 2010]:

> If the taxable income of a person at January 31, for the previous year is in excess on $100,000\$$, then the top marginal rate computed at February 28 is 50% of the total taxable income. And this provision is in force from January 1. This rule can be written as follows:

$$(Threshold^{31Jan} \rightarrow HighMarginalRate^{28Feb})^{1Jan}$$

Let us suppose that the last instalment for the salary was paid to an employee on January 4, and that it makes the total taxable income greater than the threshold stated above. We use $Threshold^{4Jan}$ to signal that the threshold of 100,000\$ has been certified on January 4. [...] So let us ask what the top marginal rate for the employee is if she lodges a tax return on January 20. [...] [From] the point of view of January 20, the top marginal rate is 50%. Suppose now that there is a change in the legislation and that the above norm is changed on February 15, and the change is that the top marginal rate is 30%.

$$(Threshold^{31Jan} \rightarrow MediumMarginalRate^{28Feb})^{15Feb}$$

In this case if the employee lodges her tax return after February 15, the top marginal rate is 30% instead of 50%. ([Governatori and Rotolo, 2010], p. 173-174)

This example shows that what can be derived depends on which rules are valid at the time when we do the derivation, especially if rules can be changed. Thus, in order to keep track of the norm changes, Governatori and Rotolo represent different versions of a legal system.

3.4.2   Back to AGM

On May 19th, 1988 a three kilometres long bridge connecting the de Ré island in the Atlantic Ocean to France was inaugurated. Among the effects of such a convenient connection was that the price per square meters on the island flared up. Suddenly, farmers whose families had been living on the island sometimes since the XVth century, found they had to pay the wealth and large fortune tax, a tax directed to individuals who own assets of high net worth. Most of those farmers are retired people with low pension, living on the products on their fields of potatoes, asparagus and vines. In order to pay the wealth and large fortune tax, some had to sell part of their fields and endangered their retirements plans. This raised serious concerns on the unexpected implications of such tax and some people advocated a change of such law.

As we have seen, one of the motivations of the AGM theory of belief revision was the study of norm change. One may also argue that some of the AGM axioms (that have been criticized in the belief revision context) appear reasonable when applied to the legal discourse. The *success* postulate for revision, for example, imposes to always accept the new input. This postulate has been heavily criticized in the belief revision literature as irrational behaviours may result from it (consider, for example, an agent who receives a stream of contradicting inputs like $\phi, \neg\phi, \phi, \neg\phi, \ldots$). The success makes however sense in the legal context, when we wish to enforce a new norm.

As we have seen in the previous subsection, the explicit temporal representation and the use of meta-rules of Governatori and Rotolo [2010]'s approach resulted in complex logics. In order to reduce such complexity, Governatori *et al.* [2013] explored three AGM-like contraction operators to remove rules, add exceptions and revise rule priorities. Similarly to Governatori and Rotolo, this approach is rooted in the legal practice. The operators and the principles are illustrated with examples taken from the Italian Constitution and real decisions taken by the Italian Constitutional Court.

Boella *et al.* [2009] (subsequently extended in [Boella *et al.*, 2016b]) also reconsidered the original inspiration of the AGM theory of belief revision as framework to evaluate the dynamics of rule-based systems. Boella *et al.* [2016b] observe that if we wish to weaken a rule-based system from which we derive too much, we can use the theory of belief base dynamics [Hansson, 1993] to select a subset of the rules as the contraction of the rule-based system.

EXAMPLE 1.1 ([Boella *et al.*, 2016b], p.274) *Consider a rule-based system consisting of the following two rules:*

1. If a then b

2. If b then c

Assume we do not want to have c in context $\{a\}$, whereas c can be derived by iteratively applying the first and the second rule. We can define rule base contraction operators that drop either the first or the second rule, or both.

However, the next example illustrates that such rule contraction operators may not be sufficient.

EXAMPLE 1.2 ([Boella *et al.*, 2016b], p.274) *Assume d is an exception to c in context a. In that case, we may want to end up with a rule base consisting of the following two rules:*

1. If $a \wedge \neg d$ then b, and

2. If b then c

or a rule base consisting of the following two rules:

1. If a then b, and

2. If $b \wedge \neg d$ then c.

In other words, in some applications, we may need to change *some of the rules. In particular, rule contraction may assume a* rule logic *which informs us that the rule 'if a then b' implies the rule 'if a ∧ ¬d then b', or that 'if b then c' implies the rule 'if b ∧ ¬d then c'.*

Thus, even if base contraction is the most straightforward and safe way to perform a contraction, it always results in a subset of the original base, which sometimes means removing too much. Take, for example $\{(a,x)\} \div (a,x) = \{\}$, where $\div$ denotes the contraction operator. Thus, under base contraction, the only result is to throw away the rule. But under AGM one can put a weaker rule. For instance, if (a,x) is the rule "If an individual owns land for more than 1.3 million Euros (a), then he must pay the wealth and large fortune tax (x)". To avoid problems as those on de Ré island, we may wish to change the law by introducing an exception, like $\{(a,x)\} \div (a,x) = \{(a \wedge b,x)\}$, where b stays for people with high income.

This was one of the motivations of Boella *et al.* [2016b]. In their abstract approach, rules are pairs (a,x) of propositional formulas and a normative system R is a set of pairs. Several logics for rules are considered by resorting to the input/output logic framework developed by Makinson and van der Torre [2000; 2003].[18]

Rules allow to derive formulas, that is, obligations and prohibitions in a normative system. The factual situation (called *context* or *input*) determines which obligations and prohibitions can be derived in a normative system. Formally, in the input/output notation: if $(a,x) \in R$ then $x \in out(R,a)$. This means that, according to the normative system R, in context a, the formula x is obligatory. The idea is that a is the input (or context) and x is the output. Of the operations defined semantically and characterized by derivation rules in Makinson and van der Torre [2000], three operations are considered in Boella *et al.* [2009; 2016b]: simple-minded, basic, and simple-minded reusable.

In order to generalize the AGM postulates for normative change, a rule set is taken to be a set of rules closed under an input/output logic. Rule expansion, rule contraction and rule revision in the input/output framework are then defined. Similarly as for the belief change case, the definition of rule expansion is unproblematic. Here, the legislator wishes to add a new rule that does not conflict with the existing ones. Rule contraction and rule revision, on the other hand, are more interesting.

AGM postulates for expansion, contraction and revision are reformulated for rule expansion, rule contraction and rule revision. It turns out that (surprisingly) the postulates for rule contraction are consistent only for some input/output logics, but not for others. On the positive side, the proof theory of rule change was shown to be closely related to the proof theory of permissions from an input/output perspective [Boella *et al.*, 2016b].

---

[18] Maranhão [2017] employs input/output logics and belief revision principles to model legal interpretation. Judicial doctrine is seen as theory change, where rules and values need to be revised to obtain a coherent system.

The translation from the AGM contraction postulates to the postulates for rule revision turned out to be more difficult. One of the difficulties was the definition of the negated input (roughly corresponding to $\neg(a,x)$) and the inconsistent set of rules in input/output logic (which would correspond to an 'incoherent' system of rules in the normative systems paradigm).

Postulates for (belief and rule) revision and (belief and rule) contraction are independent. No contraction operator appears in the revision postulates, and no revision operator appears in the postulates for contraction. Yet, the Levi identity and the Harper identity defined respectively the belief revision operator as a sequence of contraction and expansion, and the belief contraction is defined in terms of belief revision.

Using the Levi identity, rule revision was defined in terms of rule contraction. The operators so defined were shown to satisfy the AGM postulates. For the Harper identity, however, the question is still open [Boella et al., 2016b].

A similar approach to Boella et al. [2009; 2016b]'s has been proposed by Stolpe [2010]. There, AGM contractions and revision are used to define derogation and amendment of norms. In particular, the derogation operation is an AGM partial meet contraction obtained by defining a selection function for a set of norms in input/output logic. Norm revision defined via the Levi identity characterize the amendment of norms. Stolpe can thus show that derogation and amendment operators are in one-to-one correspondence with the Harper and Levi identities as inverse bijective maps.

3.5 Computational Mechanisms of Norm Change

Beside the theoretical investigations to norm change presented in the previous sections, few work exist on the computational mechanisms of norm change.

The drawback of determining norms at design time is that unforeseen situations may occur and the system cannot adapt to the new circumstances. The approach proposed by Tinnemeier et al. [2010] tackles this problem by allowing the modification of norms at runtime, so that a programmer can stipulate when and how norms can be modified. In Tinnemeier et al. [2010]'s framework norms can be modified by external agents as well as the normative framework.

The proposed norm change mechanism is system-dependent and enforcement-independent. The first principle states that who can change norms, how and when norms may be changed depends on the system. The authors justify this first principle by recalling the clause that a normative system must "specify how and in which extent the agents can modify the norms", as in the definition proposed at the first NorMAS workshop in 2005. The second principle ensures that the norm change and the norm enforcement mechanisms should be defined independently. This is to increase the readability and manageability of the program.

Two types of norm change rules are defined. The first type is used to change instances of norms without modifying the norm scheme. These rules define the circumstances under which some norm instances have to be removed to be replaced by other norm instances. The second type of rules is used to alter norm schemes. As for the first type, these rules define under which circumstances norm schemes are to be changed by retracting some norm schemes and adding others.

What happens to the instances already instantiated, when the underlying norm scheme is changed? Tinnemeier *et al.* [2010] observe that there are situations in which we want to leave the instantiated instances unchanged, and others in which it makes sense to apply the change retroactively. Thus, two types of norm scheme change rules are given. Finally, building on [Tinnemeier *et al.*, 2009], the syntax and operational semantics of the programming language are given.

Previous work on norm change at runtime includes [Bou *et al.*, 2007; Campos *et al.*, 2009]. Bou *et al.* [2007] also consider the problem of adapting a system to novel and unpredictable circumstances. To this end, they present an approach to enable normative frameworks (called "electronic institutions" in [Bou *et al.*, 2007; Campos *et al.*, 2009]) to adapt norms to agents' behaviour changes as well as to comply with institutional goals. The norm change mechanisms of Bou *et al.* [2007] allow to modify existing norms. Unlike [Tinnemeier *et al.*, 2009], new norms cannot be introduced nor can existing norms be removed. Another difference is that Bou *et al.* [2007] use a quantitative approach to represent the environment and the agents.

Campos *et al.* [2009] approached the difficulty of how to adapt a normative system to the changes of its agents' behaviour by adding situatedness and adaptation (two properties usually characterising agents) to the system. The result is a system that can make changes and that can also adapt to changes. As in Bou *et al.* [2007]'s approach, the aim is to modify agent coordination to enhance the system's performance in attaining institutional goals.

Even though Boella and van der Torre [2004]'s approach is theoretical, it shares some similarities to the works presented here. Starting from the distinction between regulative norms (that indicate what is obligatory or permitted) and constitutive (or count-as) rules (that define an activity), they use constitutive rules to create new norms as well as to define what changes the agents can introduce. As in the norm instance change rules and norm scheme change rules of Tinnemeier *et al.* [2010], constitutive and regulative rules in Boella and van der Torre [2004] are modelled as conditional rules specifying when a norm can be changed and what the consequences are.

3.6 Discussion

In this short excursus we have seen that the first formal investigations of changes in a legal code had roots in logic, namely in the AGM framework. This line of research has been reconsidered, notably in the works of Governatori and Rotolo [2010; 2013], Stolpe [2010], and Boella *et al.* [2009; 2016b], often coupled with non-classical logics such as defeasible logic or input/output. Another direction has been to follow a semantic approach inspired by dynamic logic, as in Pucella and Weissman [2004] and Aucher *et al.* [2009]. Finally, besides the theoretical investigations, some work on the computational mechanisms of norm change has been done, like Tinnemeier *et al.* [2010], Bou *et al.* [2007] and Campos *et al.* [2009].

Norm change is a fairly recent research theme in the NorMAS community. The first international workshop explicitly dedicated to the dynamics of norms was held in 2007. This observation can in part explain the lack of consensus around a common theoretical framework. But it probably does not explain it completely. Other reasons may reside in the limits of abstract frameworks like AGM, even when combined with with richer rule-based logical systems, in the difficulty to capture and distinguish norm change from changes in obligations, and again in the elusive character of legal changes in the real world. Recent developments in legal informatics may help casting light on norm dynamics. Works on legal document and knowledge management systems (like the EUNOMOS project [Boella *et al.*, 2016a]) allow, for example, to keep track of (implicit and explicit) changes in the legislation. Although these works provide some first steps in the understanding of the dynamics of normative systems, much still remains unexplored.

4 Norm Synthesis

The second theme of norm synthesis has a long-standing history but has experienced a recent revival of attention. While norm change primarily focuses on the logical implications of the modification of existing (legal) norms over time, norm synthesis puts a stronger emphasis on how (social) norms emerge and converge in the first place, and how they can be identified.

4.1 Foundations

Norm synthesis is inspired by the area of program synthesis (i.e. generating a program from a given specification [Manna and Waldinger, 1980]), but, in contrast to the former, shifts the focus to the coordination of autonomously operating agents. The specific purpose of norm synthesis is thus to identify an optimal set of norms (a normative system) to coordinate individuals' behaviours in a multi-agent system. The 'optimality' of a solution depends on the specified objectives, such as the minimal set of norms to facilitate coordination [Fitoussi and Tennenholtz, 2000].

Shoham and Tennenholtz [1992b; 1995]'s work on synthesis of social laws is considered the initial work in the area of norm synthesis. They propose a general formal model to identify a set of social laws at design time (offline) to assure the coordinated operation of structurally uniform agents. They showcase this approach by 'handcrafting' a set of social laws that guarantee collision-free coordination in a grid-based traffic scenario ('rules of the road'[19]), instead of determining action prescriptions for each possible system state. However, they also show that the automated synthesis for offline approaches is NP-hard [Shoham and Tennenholtz, 1995], challenging the generalizable application. Onn and Tennenholtz [1997] propose a general solution for the synthesis problem for scenarios that can be represented as biconnected graphs by reducing synthesis to a graph routing problem. Fitoussi and Tennenholtz [2000] further introduce qualitative characteristics for synthesized social laws, such as their *Minimality* and *Simplicity*. As alluded to before, minimal social laws seek to specify fewest possible restrictions on agents' behaviours, thus giving individuals the greatest possible autonomy, while maintaining coordination in the overall system. An extremely restrictive social law would effectively prescribe any action an agent could take in any given situation (e.g. to walk on the right side of a footpath in a given direction, or even more restrictive, prescribing specific navigation routes between different locations), thus removing any form of autonomy on the part of the agent. A minimal social law (e.g. not to step on the road), in contrast, would retain the agent's ability to pursue its own goals, as long as it is compatible with the system objectives (e.g. avoiding collisions between cars and pedestrians). In a more recent approach, Christelis and Rovatsos [2009]'s work on automated offline norm synthesis addresses the complexity problem by identifying prohibitive states in incomplete state specifications that are generalized across the entire state space. It is important to note that these early approaches to norm synthesis do not consider or tolerate any form of violation; unlike most subsequent work, their conceptions of social laws describe hard constraints agents cannot forego.

The shift towards refined norm interpretations that emphasizes the interactionist over legal perspective (and thus regulation over regimentation) [Boella *et al.*, 2008] has stimulated a differentiated treatment of rewards and sanctions as mechanisms of social enforcement. This sociologically-inspired norm perspective drove the exploration of associated influence factors (such as memory and connectivity), along with a movement from *offline* to *online* norm synthesis, resulting in two subfields. *Convention/Norm Emergence* (which we will differentiate later) emphasize mechanisms that influence the convergence on norms or conventions, whereas work we cluster under the label *Identification* concentrates on the mechanics of detecting and synthesising norms in the first place. The latter can further be subdivided into approaches that rely on a centralized or decentralized operation, that is, approaches that use a central entity to synthesize norms, or delegate the generalization and integration of identified norms to the agents themselves. Figure 7 provides a schematic overview of the outlined structure of the research field. Overall, the subfields of norm synthesis cover the

---

[19]This *de facto* reference scenario has been adopted and refined in large parts of subsequent work on norm synthesis.

notion of norms in the broad sense (i.e. as institutions), ranging from self-enforcing conventions via socially enforced norms to centrally enforced social laws or rules. In the following, we will discuss selected contributions to the area of norm synthesis, with a particular focus on approaches that emphasize the detection and identification of norms.

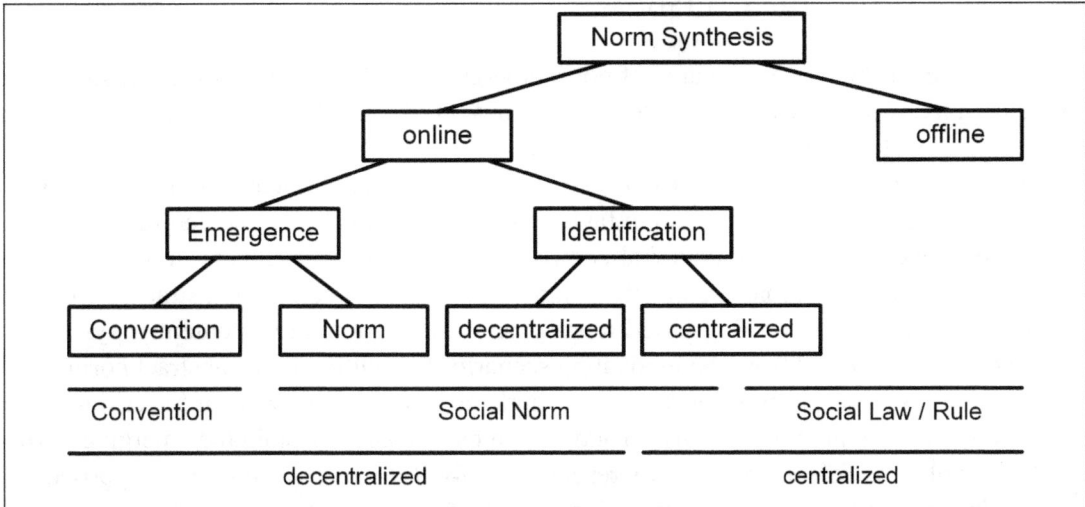

Figure 7: Taxonomy of Norm Synthesis Approaches

4.2 Synthesis as Norm/Convention Emergence

Research efforts in the area of *norm emergence* put particular concentration on an understanding of the contextual conditions and mechanisms that bring norms about, including their distributed nature. Instead of relying on a centralized entity to determine norms a priori or embedding hard-coded (offline designed) norms into individuals, norm emergence affords the decentralized collaboration of agents to converge on commonly accepted social norms.

Explored mechanisms include:

- Memory size (e.g. Villatoro *et al.* [2009])

- Network topologies and dynamics of relationships (e.g. Savarimuthu *et al.* [2009], Villatoro *et al.* [2009], Sen and Sen [2010], Sugawara [2011], Villatoro *et al.* [2013])

- Clusters (e.g. Pujol *et al.* [2005])

- Interaction-based social learning (e.g. Sen and Airiau [2007], Mukherjee *et al.* [2007; 2008], Airiau *et al.* [2014])

- Lying (e.g. Savarimuthu *et al.* [2011])

- Dynamic sanctions (e.g. Mahmoud *et al.* [2012; 2015])

- Hierarchical structures with varying levels of influence (e.g. Franks *et al.* [2013; 2014], Yu *et al.* [2013; 2015])

Further contributions in the area of norm emergence include algorithms for distributed decision-making to arrive at a shared lexicon [Salazar *et al.*, 2010] or shared sets of tags [Griffiths and Luck, 2010].

The decentralized operation of norm emergence places an emphasis on larger number of agents and their direct interaction in favour of cognitive capability and central coordination. Consequently, the computational complexity of individual agents is limited and the applied norm representations are mostly abstract in the form of converging strategy choices in coordination games or string-based representations; the normative content is symbolic and can only be inferred from the motivating scenario. In addition to the abstract normative content, in most cases, agents converge on a single norm (with exception of Savarimuthu *et al.* [2009] and Sen and Sen [2010]). In addition, most scenarios sustain the emerging norm without explicit enforcement, thus representing *self-enforcing conventions* as opposed to *externally enforced social norms*, affording the differentiation into *Convention Emergence* and *Norm Emergence*.

Following the exploration of the emergence strand of norm synthesis, we will turn to the identification strand that captures norm synthesis processes in a narrow sense, primarily focusing on the detection, identification, and integration of norms into consistent normative systems.

4.3   Synthesis as Identification

Work that identifies and synthesizes norms at runtime can be differentiated into centralized approaches, which interpret norm synthesis in the original spirit of identifying centrally managed system-wide norms, and decentralized ones that analyze the inception of norms from a bottom-up perspective.

A series of centralized online norm synthesis approaches that follow the tradition of Shoham and Tennenholtz has been spearheaded by Morales *et al.*. In their work, Morales *et al.* [2013] propose the *Intelligent Robust Norm Synthesis* mechanism dubbed IRON in an adapted version of the grid-based 'rules of the road' scenario originally introduced by Shoham and Tennenholtz that focuses on coordination in traffic junctions. Agents have a limited observational range and move in travel direction, unless constrained by imposed norms. IRON continually monitors traffic participants' behaviour. When detecting collisions, IRON identifies the underlying conditions (e.g. car approaching from the right) and introduces a norm that prevents a similar event from reoccurring (e.g. by introducing an obligation to stop whenever facing a car to one's right). These centrally generated and managed norms (which make those effectively rules or social laws) are imposed upon all traffic participants, thus progressively moving towards a stable collision-free normative system.

To prevent overregulation from introducing too many specific norms based on individual observations, IRON attempts to *generalize* norms based on their shared preconditions by selectively ignoring a specific norm's partial precondition. The generalized norm is evaluated at runtime by detecting eventual recurring collisions, in which case the original specific norms are deemed relevant and are reinstated. To determine the *effectiveness* of given norms, IRON further monitors their activation, and ascribes frequently applied norms higher effectiveness. To identify *necessary* norms, Morales *et al.* [2013] (unlike Shoham and Tennenholtz [1995]'s social law approach) make use of the agents' ability to violate norms, which enables IRON to identify imposed norms that are actually necessary to maintain coordination and remove unnecessary ones (i.e. norms whose violation does not produce collisions).

Morales *et al.* [2014] successively introduce further iterations of their approach (dubbed SIMON) that consider structural diversity of norm participants (e.g. by introducing emergency vehicles) and refined mechanisms for norm generalization with specific focus on minimizing the necessary simulation runtime to produce a collision-free normative system Morales *et al.* [2014; 2015c] . Their following system iteration, LION [Morales *et al.*, 2015b], includes the focus on the identification of semantic relationships between norms, so as to produce fewer, more general norms (liberal norms) that maximize the norm participants' autonomy.

This series of works on norm synthesis highlights the advantages of centralized approaches not only to identify norms, but to integrate those. In this interpretation, synthesis involves an explicit analytical effort to integrate individual norms into a coherent normative system, producing semantically meaningful complex coordination outcomes, beyond a coordinated strategy choice as observed in most norm emergence approaches. Consequently, a comprehensive approach to norm synthesis captures life cycle processes that include identification, as well as internalization and forgetting of norms, thus covering processes that are associated with the evolution of norms over time (see Section 2.6). Processes such as spreading and enforcement, characteristically associated with the work on norm emergence, are secondary.

Riveret *et al.* [2014]'s transfiguration approach takes an incremental step towards decentralized systems by endowing individual agents with learning capabilities enabling them to infer behavioural prescriptions from stochastic games. Being grounded in the field of computational justice, their approach marries bottom-up dynamics (transfiguration of experience into prescriptions) with notions of self-governance by means of collective action (voting). The voting process is initiated once all agents have submitted their inferred (and preferred) prescriptions, the most common of which is put forth as a motion. Agents are then invited to vote based on the perceived purposefulness of the prescription content, which is abstractly represented using a notion of global and individually perceived *potential*. Since the purpose of the voting process (in the spirit of self-governance) is to promote globally useful prescriptions, the agents decide probabilistically based on the alignment of the candidate prescription's individual and global potential. Once adopted, the prescription becomes a self-imposed rule of that society.

This work emphasizes the computational representation of social processes that enable self-governance by retaining high levels of decisional autonomy on the part of the society members, while abstractly providing centralized decision-making and enforcement inspired by real societies. Beyond the conceptual integration of bottom-up and top-down governance processes, this contribution emphasizes the efficiency benefits associated with centrally co-ordinated collective decision-making.

Contributions that shift the perspective away from approaches that emphasize effective coordination towards individual-centric operations can be captured under the umbrella of *decentralized online norm synthesis*. In addition to the focus on the individual as an entity of concern, in principle these approaches lend themselves well for explorative scenarios with a broader (if not open) range of actions than used in the centralized coordination scenarios. Research efforts related to this cluster include Andrighetto *et al.* [2007; 2010] as well as Savarimuthu *et al.* [2010b; 2013a]. We will not discuss these works in great detail at this stage as we covered those in the context of norm creation in Hollander and Wu [2011b]'s life cycle model (see Section 2.3). Instead, we will concentrate on contributions that treat norm synthesis as a holistic process involving multiple life cycle processes.

An important work in this area is Savarimuthu *et al.* [2013b]'s work on norm recommendation. Their approach is motivated by the identification and recommendation of an existing system's norms to newcomers, which can operate in a centralized or decentralized fashion. Their system combines norm identification, classification and life cycle stage detection in order to recommend the existence and relevance of observed norms. The initial step of norm detection operates on a continuous stream of events by identifying recurring event episodes that are terminated with a sanction signal. The algorithm collects event episodes of varying window sizes in order to establish the subset of actions that provoke a sanction signal and identifies those as candidate norms. In the second step, norms are classified with respect to their salience. For this purpose, the mechanism tracks both the invocation of actions contained in the candidate norms as well as the frequency of punishments as a response to action activation. By ranking these measures, the mechanism classifies norms by salience, where the existence of punishment is indicative of higher levels of salience, as opposed to mere action activation. A further step emphasizes the long-term perspective and attempts to identify a norm's life cycle stage (*life stage*), with possible stages being emerging, growing, maturing, declining, and decaying. The system monitors norms' punishment probabilities over time and evaluates those with respect to given successive thresholds associated with emergence (frequency of activation) and growth, based on which it infers the life stage. For example, norms that have experienced an increase in punishment between two time intervals but remain between the emergence and growth thresholds, are considered growing. The system uses heuristics that use the established measures for salience and life stage as an input to *recommend* the existence of a given norm.

Similar to Morales *et al.* [2015a]'s works, Savarimuthu *et al.* [2013b]'s synthesis approach allows the identification of multiple norms, along with a quantitative measure of salience that is comparable with Morales *et al.* [2013]'s notion of effectiveness and necessity. Savarimuthu *et al.* [2013b]'s approach further includes a systematic classification of norms with respect to their life cycle stage, thus emphasising the long-term perspective. However, unlike Morales *et al.* [2015a], this work relies on an abstract string-based norm representation and does not consider semantic relationships between norms, thus preventing operations such as generalization or substitution of norms.

The final approach we present under the umbrella of norm synthesis takes an intermediate stance by operating decentralized while maintaining meaningful norm representations. Frantz *et al.* [2014c; 2015] propose a norm generalization approach that operates on individual observations. At its core, this approach is motivated by individuals' tendency to subconsciously develop stereotypes as decision-making shortcuts they can use when encountering unknown interaction partners. To facilitate this generalization, the mechanism relies on uniform structural representations of actors, actions and norms based on Nested ADICO (nADICO) [Frantz *et al.*, 2013; Frantz *et al.*, 2015], a rule-based norm representation that builds on the *Grammar of Institutions* [Crawford and Ostrom, 1995] and affords the explicit representation of structural institutional regress [Frantz, 2015], i.e. the nested interdependency of sanctions and corresponding metanorms. As a first step, observations are aggregated based on shared observable attributes as well as subsets thereof (higher generalization levels), forming the basis to synthesize descriptive norms (or conventions) the observer attributes to observed groups of individuals. To infer injunctive norms from observations, individuals further track corresponding reactions to ascribe the generalized action sequences normative character and interpret the generalized reactions as social consequences (i.e. rewards or sanctions). The frequency and intensity of observations indicate a norm's salience by mapping it onto a continuous deontics conception (*Dynamic Deontics* [Frantz *et al.*, 2014a]) that spans from prohibition via permission to obligation, the *deontic range* of which is unique for each agent and determined by its previous experience. In addition to the extremal cases, this concept introduces intermediate stages along this continuum (e.g. obligations that are omissible and can be exceptionally foregone), a principle that is used to reflect the subjectively perceived priority of a given norm, and implicitly solves potential norm conflicts.

In contrast to the approach by Morales *et al.* [2015a], this work does not solve a specific coordination problem, but introduces a fully decentralized approach to understand agents' behaviours by inspecting their *subjective understanding* of a scenario's normative content, thus shifting it into closer proximity to emergence-based approaches. Similar to Morales *et al.* [2015a] (but unlike Savarimuthu *et al.* [2013b]), this approach uses a comprehensive human-readable norm representation (as *institutional statements*) and allows the identification of norm relationships by generalizing individual observations. The uniform norm representation further permits the analysis on arbitrary social aggregation levels (e.g. group, society).

Table 3 provides a chronological overview of all identified norm synthesis approaches based on the characteristics introduced at the beginning of this subsection (see Figure 7), including the nature of norm (convention, norm, rule, social law), central coordination and the ability to produce or identify multiple norms.

Contribution	Institution Type	Centralized	Offline	Single Norm
Shoham and Tennenholtz [1995]	Social Law	yes	yes	no
Pujol *et al.* [2005]	Convention	no	no	yes
Sen and Airiau [2007]	Convention	no	no	yes
Savarimuthu *et al.* [2007; 2008a]	Norm	no	no	no
Mukherjee *et al.* [2007; 2008]	Convention	no	no	yes
Christelis and Rovatsos [2009]	Social Law	yes	yes	no
Villatoro *et al.* [2009]	Convention	no	no	yes
Urbano *et al.* [2009]	Convention	no	no	yes
Sen and Sen [2010]	Convention	no	no	yes
Griffiths and Luck [2010]	Norm	no	no	no
Sugawara [2011]	Convention	no	no	no
Mahmoud *et al.* [2012]	Norm	no	no	yes
Morales *et al.* [2013]	Social Law	yes	no	no
Franks *et al.* [2013]	Convention	no	no	yes
Villatoro *et al.* [2013]	Convention	no	no	yes
Savarimuthu *et al.* [2013b]	Norm	both	no	no
Mihaylov *et al.* [2014]	Convention	no	no	yes
Airiau *et al.* [2014]	Convention	no	no	yes
Morales *et al.* [2014]	Social Law	yes	no	no
Riveret *et al.* [2014]	Norm / Rule	yes	no	no
Frantz *et al.* [2014c; 2015]	Norm	no	no	no
Morales *et al.* [2015b]	Social Law	yes	no	no
Mahmoud *et al.* [2015]	Norm	no	no	yes

Table 3: Overview of Norm Synthesis Approaches

4.4 Contextualization with the General Norm Life Cycle Model

At the current stage, norm synthesis presents itself as a diverse field that is driven by varying objectives. Apart from the historical separation into offline and online approaches, we can identify a cluster of existing approaches that either concentrate on the:

- Investigation of factors and circumstances that promote norm adoption (emphasizing macro-level outcomes), or

- Mechanisms for the runtime identification, generalization, implementation, and integration with established norms (emphasizing micro-level mechanisms).

Relating these approaches to individual life cycle processes of the general norm life cycle model (see Section 2.7) as shown in Figure 8, we can observe that emergence-based approaches emphasize spreading/transmission mechanisms (e.g. type and dynamic nature of network topologies, hierarchical structures, social learning, memory size) along enforcement characteristics (e.g. sanctioning, lying).

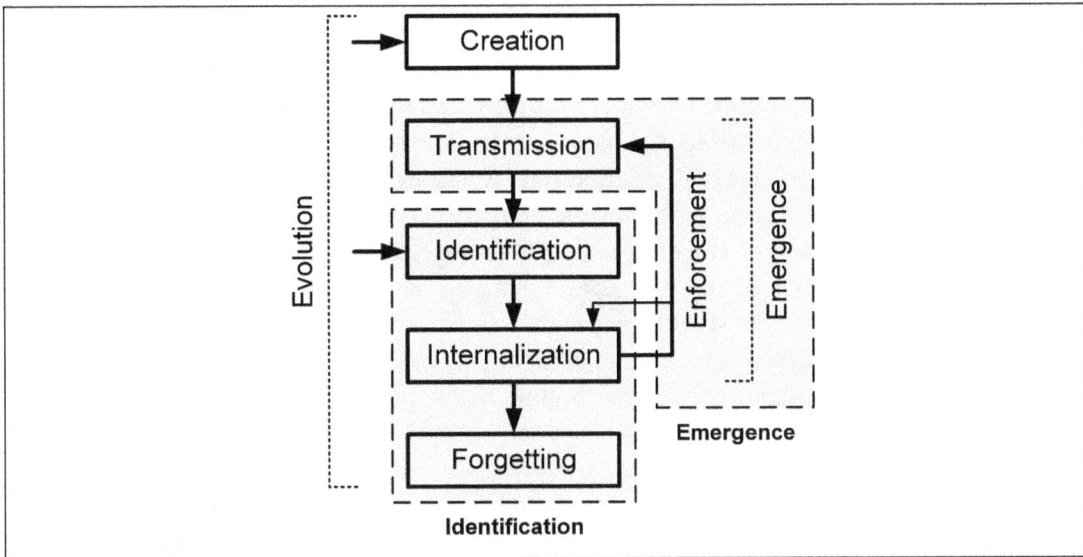

Figure 8: Norm Synthesis Approaches and Related Life Cycle Processes

The second group of mechanisms emphasize the detection and identification of existing norms. Deductive tasks for the generalization of comprehensive normative systems are related to a complex norm internalization process (such as the one conceptualized by Hollander and Wu [2011b]), since it represents a composite process that merges newly discovered norms and existing sets of norms, which requires the ability to modify, generalize and integrate norms. The synthesis of normative systems further relies on the ability to discard or forget norms.

Despite the comprehensive coverage of different life cycle stages, the review of existing approaches indicates gaps. An important central topic that has found limited explicit attention in current approaches is the detection of norm conflicts, an aspect with a strong relation to the norm internalization process. Riveret *et al.* [2014], as well as Savarimuthu *et al.* [2013b], treat norms independently without considering their relationship to existing norms. Frantz *et al.* [2015] and Morales *et al.* [2015a] include generalization processes and mechanisms to accommodate conflicting or competing norms, but only Morales *et al.* [2015b] perform explicit detection of norm relationships such as complementarity and substitutability. An area that has found recent attention is the focus on *dynamic normative systems* [Huang *et al.*, 2016] in which the normative environment itself is not considered static, but changes over time, and thus requires agents to revise their normative understanding in order to accommodate those changes. Initial work by Huang *et al.* [2016] analyzes the associated complexity of norm recognition and synthesis.

4.5 Discussion of Challenges and Future Directions

In this section, we provided a comprehensive discussion of the historical roots of norm synthesis, the shifts from offline to online synthesis, and the subsequent differentiation into more implicit emergence-focused and more explicit identification-centric approaches. We further discussed a set of relevant contributions to the latter identification strand of norm synthesis. However, apart from surveying the field, this comprehensive overview of the area of norm synthesis allows us to identify areas which we believe deserve further attention.

Reviewing the strands of (online) norm synthesis, an outstanding development is the systematic integration of both strands by enriching emergence-based approaches with richer micro-level architectures that incorporate components of identification-based mechanisms. For identification-based approaches, this implies a stronger focus on generalizable representations of norms and social structures beyond specific scenarios. The marriage of both strands provides a basis for more realistic representations of social scenarios, with *emergence* sponsoring the insight on how to structure interaction in social environments, and *identification* providing mechanisms to develop complex, yet consistent normative systems as we encounter them in the real world.

We further believe that the exploration of dynamic normative systems represents an important research direction if we aim towards the use of norm synthesis in real-world applications (e.g. robotics). It further has the potential to link the theoretical contributions developed in the area of norm change, e.g. modelling changes in legal systems (as discussed in Section 3), with the mechanisms that facilitate the identification, generalization and integration of corresponding operational norms developed in the area of norm synthesis.

Looking beyond the scope of contemporary work, an important challenge for the successful adoption of norm synthesis is the identification and development of application domains that enable the use of these techniques in realistic scenarios, both involving the extent and complexity of available data. In this context, a challenge that all contemporary approaches to norm synthesis share is their operation on structured data. Making unstructured, noisy or semi-structured data (such as found in big data) accessible under consideration of the complexity limitations of current norm synthesis approaches will increase its relevance for real-world applications. Specific examples include the automated the extraction of norms from large and diverse real-world data corpi, as well as performing online norm synthesis, e.g. for the ad hoc inference of normative understanding in the context of robotics or digital assistants.

5 Summary, Conclusions and Outlook

In this article, we have provided an overview of the contemporary perspective on norm dynamics, with a particular focus on norm change and norm synthesis as important active research fields in multi-agent systems.

The research around norm change (Section 3) has resulted in a comprehensive exploration of logical challenges associated with the representation of changing social and legal norms, such as temporal implications of changing laws and an adequate formal translation of the notion of an incoherent normative system. At this stage, the relatively young but promising field has yet to find a shared consensus on the theoretical foundation to provide the platform for the systematic application of its contributions in the context of normative multi-agent systems as well as other disciplines.

Research in the area of norm synthesis (Section 4) concentrates on the analysis of factors that contribute to emerging norms (norm emergence) as well mechanisms to detect existing norms (norm identification). This field has experienced a revival with the recent focus on the synthesis of normative systems at runtime (online) – as opposed to the traditional offline approach. In addition, the field features an increasing number of approaches that favour decentralized over centralized approaches or combine both approaches and use social choice mechanisms for the integration of bottom-up and top-down perspectives on norm synthesis.

To understand the developments in both fields, we initially presented an overview of approaches that define the norm life cycles (Section 2), while providing an overview of the contemporary state of current contributions associated with individual life cycle processes. We further systematically compared the surveyed life cycles based on involved processes and norm characteristics, while identifying abstract phases of the norm life cycle. From this analysis, we extracted the essential processes and integrated those in a *general norm life cycle model* that reflects the contemporary view on norm emergence. The refined model resolves terminological and conceptual inconsistencies/omissions identified in the existing life cycle models. It further suggests that external influence factors can lead to norm modification throughout all stages of the norm life cycle, and, unlike earlier models, distinguishes between normative processes and associated phenomena.

Since this article specifically concentrates on the *modeling of norm dynamics*, we do not capture the wider technical and philosophical implications of norm dynamics, such as the dealing with normative conflicts and violations (see article 'Modeling Normative Conflicts in Multi-Agent Systems' in this volume), aspects of norm autonomy (see Verhagen [2000]), and the role of trust for the functioning of norms (see Andrighetto *et al.* [2013]).

Surveying individual contributions to the field of NorMAS in general – and to the areas of norm change and synthesis in particular – we can observe a tendency to apply richer norm conceptions that span across multiple norm life cycle processes. As a result, developed systems produce increasingly dynamic behaviour. This includes a) the identification of norms at runtime, b) the change of norms over time, and c) their potential decay and substitution.

These observations highlight an important progression for the wider discipline, since it positions the current development on the roadmap laid out in the 2007 Dagstuhl NorMAS workshop that identified five levels in the development of normative multi-agent systems (see Boella *et al.* [2008]):

- Level 1 – Off-line designed norms

- Level 2 – Explicit norm representations that can be used for communication and negotiation

- Level 3 – Runtime addition, removal and modification of norms

- Level 4 – Embeddedness in social reality

- Level 5 – Development of moral reality

The first three levels are undisputed – the shift towards dynamic creation (Level 3) is reflected in numerous contributions to the field. However, the ability of agents to identify and synthesize norms in their social environment at runtime, the ability to engage in social choice processes, as well as agents' compliance in dynamic normative systems provide the basis to make agents active participants in shaping social reality, and thus moves them closer to the fourth development level (without discussing the associated challenges at this stage – for details see Boella *et al.* [2008]).

Fundamentally, this integration of normative concepts in social reality cannot be dissociated from the consideration of ethical and moral concerns as suggested for the last level – the development of moral reality by assuming moral agency. This resonates with contemporary developments, such as the productive use of autonomous cars, increasing automation of the workforce via robotics, decentralisation of autonomy (e.g. in distributed ledger technology), along with the revived societal debates around the impact of artificial intelligence (e.g. recall the debates around universal base income). This necessity to address the embeddedness in social reality and moral reality at the same time is reflected in calls for future research directions in artificial intelligence (e.g. Russell *et al.* [2015]) and visible in initial contributions towards that end (e.g. Conitzer *et al.* [2017]).

These general AI challenges provide a unique opportunity for the interdisciplinary field of normative multi-agent systems. This field studies the very dynamics that allow systems to address fuzzy and complex problems conventional rule-based systems are not prepared to deal with. It does so by exploiting two central features of norms, a) their adaptiveness towards changing social and technological environments, and b) their innate scalability based on their decentralized operation. Independent of the application domain, this leaves us researchers with the task to foster and establish an interdisciplinary operationalisation of norms as dynamic decentralized coordination mechanisms. This, in consequence, makes norm dynamics an integral component for the modelling of realistic social behaviour within and beyond normative multi-agent systems.

References

[Airiau *et al.*, 2014] S. Airiau, S. Sen, and D. Villatoro. Emergence of conventions through social learning. *Autonomous Agents and Multi-Agent Systems*, 28(5):779–804, 2014.

[Alchourrón and Makinson, 1981] C. E. Alchourrón and D. Makinson. Hierarchies of regulations and their logic. 1981. in [?] 125–148.

[Alchourrón and Makinson, 1982] C. E. Alchourrón and D. Makinson. On the logic of theory change: Contraction functions and their associated revision functions. *Theoria*, 48:14–37, 1982.

[Alchourrón *et al.*, 1985] C. E. Alchourrón, P. Gärdenfors, and D. Makinson. On the logic of theory change: Partial meet contraction and revision functions. *J. Symb. Log.*, 50(2):510–530, 1985.

[Andreoni, 1989] J. Andreoni. Giving with impure altruism: Applications to charity and ricardian equivalence. *Journal of Political Economy*, 97(6):1447–1458, 1989.

[Andrighetto *et al.*, 2007] G. Andrighetto, M. Campenni, R. Conte, and M. Paolucci. On the immergence of norms: A normative agent architecture. In *Proceedings of the AAAI Symposium, Social and Organizational Aspects of Intelligence*, Washington DC, 2007.

[Andrighetto *et al.*, 2010] G. Andrighetto, M. Campenni, F. Cecconi, and R. Conte. The complex loop of norm emergence: A simulation model. In K. Takadama, C. Cioffi-Revilla, and G. Deffuant, editors, *Simulating Interacting Agents and Social Phenomena*, pages 19–35. Springer, Berlin, 2010.

[Andrighetto *et al.*, 2013] G. Andrighetto, C. Castelfranchi, E. Mayor, J. McBreen, M. Lopez-Sanchez, and S. Parsons. (Social) norm dynamics. In G. Andrighetto, G. Governatori, P. Noriega, and L. van der Torre, editors, *Normative Multi-Agent Systems. Vol. 4 of Dagstuhl Follow-Ups*, pages 135–170, 2013.

[Aucher *et al.*, 2009] G. Aucher, D. Grossi, A. Herzig, and E. Lorini. Dynamic context logic. In X. He, J. Horty, and E. Pacuit, editors, *Logic, Rationality, and Interaction: Second International Workshop, LORI 2009, Chongqing, China, October 8-11, 2009. Proceedings*, pages 15–26, Berlin, Heidelberg, 2009. Springer Berlin Heidelberg.

[Axelrod, 1986] R. Axelrod. An evolutionary approach to norms. *The American Political Science Review*, 80(4):1095–1111, 1986.

[Baldwin, 1971] D. A. Baldwin. The power of positive sanctions. *World Politics*, 24(1):19Ű38, 1971.

[Bandura, 1977] A. Bandura. *Social Learning Theory*. General Learning Press, New York (NY), 1977.

[Barabási and Albert, 1999] A.-L. Barabási and R. Albert. Emergence of scaling in random networks. *Science*, 286(5439):509–512, October 1999.

[Beheshti *et al.*, 2015] R. Beheshti, A. M. Ali, and G. Sukthankar. Cognitive social learners: An architecture for modeling normative behavior. In *Proceedings of the Twenty-Ninth AAAI Conference on Artificial Intelligence*, AAAI'15, pages 2017–2023. AAAI Press, 2015.

[Bianco and Bates, 1990] W. T. Bianco and R. Bates. Cooperation by design: Leadership, structure, and collective dilemmas. *American Political Science Review*, 84(1):133–147, 1990.

[Bicchieri, 2006] C. Bicchieri. *The Grammar of Society: The Nature and Dynamics of Social Norms*. Cambridge University Press, New York, 2006.

[Boella and van der Torre, 2004] G. Boella and L. van der Torre. Regulative and constitutive norms in normative multiagent systems. In *IN PROCS. OF KRÕ04*, pages 255–265. AAAI Press, 2004.

[Boella *et al.*, 2006] G. Boella, L. van der Torre, and H. Verhagen. Introduction to normative multiagent systems. *Computation and Mathematical Organizational Theory*, 12(2-3):71–79, 2006.

[Boella *et al.*, 2008] G. Boella, L. van der Torre, and H. Verhagen. Introduction to special issue on normative multiagent systems. *Autonomous Agents and Multi-Agent Systems*, 17(1):1–10, 2008.

[Boella *et al.*, 2009] G. Boella, G. Pigozzi, and L. van der Torre. Normative framework for normative system change. In *8th International Joint Conference on Autonomous Agents and Multiagent Systems (AAMAS 2009), Budapest, Hungary, May 10-15, 2009, Volume 1*, pages 169–176, 2009.

[Boella *et al.*, 2016a] G. Boella, L. D. Caro, L. Humphreys, L. Robaldo, P. Rossi, and L. vanäder Torre. Eunomos, a legal document and knowledge management system for the web to provide relevant, reliable and up-to-date information on the law. *Artificial Intelligence and Law*, 24(3):245–283, 2016.

[Boella *et al.*, 2016b] G. Boella, G. Pigozzi, and L. van der Torre. AGM contraction and revision of rules. *Journal of Logic, Language and Information*, 25(3-4):273–297, 2016.

[Boman, 1999] M. Boman. Norms in artificial decision making. *Artificial Intelligence and Law*, 7(1):17–35, 1999.

[Bou *et al.*, 2007] E. Bou, M. López-Sánchez, and J. A. Rodríguez-Aguilar. Adaptation of autonomic electronic institutions through norms and institutional agents. In G. M. P. O'Hare, A. Ricci, M. J. O'Grady, and O. Dikenelli, editors, *Engineering Societies in the Agents World VII: 7th International Workshop, ESAW 2006 Dublin, Ireland, September 6-8, 2006 Revised Selected and Invited Papers*, pages 300–319, Berlin, Heidelberg, 2007. Springer Berlin Heidelberg.

[Bourdieu, 1977] P. Bourdieu. *An Outline of a Theory of Practice*. Cambridge University Press, London, 1977.

[Boyd and Richerson, 1985] R. Boyd and P. Richerson. *Culture and the Evolutionary Process*. University of Chicago Press, Chicago (IL), 1985.

[Boyd and Richerson, 2005] R. Boyd and P. Richerson. *The Origin and Evolution of Cultures*. Oxford University Press, New York (NY), 2005.

[Bratman, 1987] M. Bratman. *Intentions, Plans, and Practical Reason*. Harvard University Press, Cambridge (MA), 1987.

[Bravo *et al.*, 2012] G. Bravo, F. Squazzoni, and R. Boero. Trust and partner selection in social networks: An experimentally grounded model. *Social Networks*, 34(4):481–492, 2012.

[Broersen *et al.*, 2001] J. Broersen, M. Dastani, J. Hulstijn, Z. Huang, and L. van der Torre. The BOID architecture: Conflicts between beliefs, obligations, intentions and desires. In *Proceedings of the Fifth International Conference on Autonomous Agents*, AGENTS '01, pages 9–16, New York, NY, USA, 2001. ACM.

[Broersen *et al.*, 2002] J. Broersen, M. Dastani, J. Hulstijn, and L. van der Torre. Goal generation in the BOID architecture. *Cognitive Science Quarterly*, 2(3-4):428–447, 2002.

[Caldas and Coelho, 1999] J. C. Caldas and H. Coelho. The origin of institutions: socio-economic processes, choice, norms and conventions. *Journal of Artificial Societies and Social Simulation*, 2(2):1, 1999.

[Campenni *et al.*, 2009] M. Campenni, G. Andrighetto, F. Cecconi, and R. Conte. Normal = Normative? The role of intelligent agents in norm innovation. *Mind & Society*, 8(2):153–172, 2009.

[Campos *et al.*, 2009] J. Campos, M. López-Sánchez, J. A. Rodríguez-Aguilar, and M. Esteva. Formalising situatedness and adaptation in electronic institutions. In J. F. Hübner, E. Matson, O. Boissier, and V. Dignum, editors, *Coordination, Organizations, Institutions and Norms in Agent Systems IV : COIN 2008 International Workshops, COIN@AAMAS 2008, Estoril, Portugal, May 12, 2008. COIN@AAAI 2008, Chicago, USA, July 14, 2008. Revised Selected Papers*, pages 126–139, Berlin, Heidelberg, 2009. Springer Berlin Heidelberg.

[Castelfranchi *et al.*, 1998] C. Castelfranchi, R. Conte, and M. Paolucci. Normative reputation and the costs of compliance. *Journal of Artificial Societies and Social Simulation*, 1(3):3, 1998.

[Chalub *et al.*, 2006] F. Chalub, F. Santos, and J. Pacheco. The evolution of norms. *Journal of Theoretical Biology*, 241(2):233–240, 2006.

[Checkel, 1998] J. Checkel. The constructivist turn in international relations theory. *World Politics*, 50(2):324–348, 1998.

[Christelis and Rovatsos, 2009] G. Christelis and M. Rovatsos. Automated norm synthesis in an agent-based planning environment. In *Proceedings of The 8th International Conference on Autonomous Agents and Multiagent Systems - Volume 1*, AAMAS '09, pages 161–168, Richland, SC, 2009. International Foundation for Autonomous Agents and Multiagent Systems.

[Conitzer *et al.*, 2017] V. Conitzer, W. Sinnott-Armstrong, J. S. Borg, Y. Deng, and M. Kramer. Moral decision making frameworks for artificial intelligence, 2017.

[Conte and Castelfranchi, 1995a] R. Conte and C. Castelfranchi. *Cognitive and Social Action*. UCL Press, London, 1995.

[Conte and Castelfranchi, 1995b] R. Conte and C. Castelfranchi. Understanding the effects of norms in social groups through simulation. In N. Gilbert and R. Conte, editors, *Artificial Societies: The Computer Simulation of Social Life*, pages 252–267. UCL Press, London, 1995.

[Crawford and Ostrom, 1995] S. E. Crawford and E. Ostrom. A Grammar of Institutions. *The American Political Science Review*, 89(3):582–600, September 1995.

[Delgado *et al.*, 2003] J. Delgado, J. M. Pujol, and R. Sangüesa. Emergence of coordination in scale-free networks. *Web Intelligence and Agent Systems*, 1(2):131–138, April 2003.

[Delgado, 2002] J. Delgado. Emergence of social conventions in complex networks. *Artificial Intelligence*, 141(1–2):171–185, 2002.

[Eguia, 2011] J. X. Eguia. *A Theory of Discrimination and Assimilation*. New York University Press, New York (NY), 2011.

[Ehrlich and Levin, 2005] P. R. Ehrlich and S. A. Levin. The evolution of norms. *PLoS Biology*, 3(6), 06 2005.

[Elster, 1989] J. Elster. Social norms and economic theory. *Journal of Economic Perspectives*, 3(4):99–117, 1989.

[Epstein, 2001] J. M. Epstein. Learning to be thoughtless: Social norms and individual computation. *Computational Economics*, 18(1):9–24, 2001.

[Erdős and Rényi, 1959] P. Erdős and A. Rényi. On random graphs i. *Publicationes Mathematicae*, 6:290–297, 1959.

[Fermé and Hansson, 1999] E. L. Fermé and S. O. Hansson. Selective revision. *Studia Logica: An International Journal for Symbolic Logic*, 63(3):331–342, 1999.

[Finnemore and Sikkink, 1998] M. Finnemore and K. Sikkink. International norm dynamics and political change. *International Organization*, 52(4):887–917, 1998.

[Fitoussi and Tennenholtz, 2000] D. Fitoussi and M. Tennenholtz. Choosing social laws for multi-agent systems: Minimality and simplicity. *Artificial Intelligence*, 119(1–2):61–101, 2000.

[Fix *et al.*, 2006] J. Fix, C. von Scheve, and D. Moldt. Emotion-based norm enforcement and maintenance in multi-agent systems: Foundations and Petri net modeling. In H. Nakashima, M. P. Wellman, G. Weiss, and P. Stone, editors, *Proceedings of the 5th International Joint Conference on Autonomous Agents and Multi-Agent Systems*, AAMAS '06, pages 105–107, New York (NY), 2006. ACM Press.

[Flentge *et al.*, 2001] F. Flentge, D. Polani, and T. Uthmann. Modelling the emergence of possession norms using memes. *Journal of Artificial Societies and Social Simulation*, 4(4):3, 2001.

[Franks *et al.*, 2013] H. Franks, N. Griffiths, and S. Anand. Learning influence in complex social networks. In Ito, Jonker, Gini, and Shehory, editors, *Proceedings of the 12th International Conference on Autonomous Agents and Multi-Agent Systems*, AAMAS '13, pages 447–454, 2013.

[Franks *et al.*, 2014] H. Franks, N. Griffiths, and S. S. Anand. Learning agent influence in MAS with complex social networks. *Autonomous Agents and Multi-Agent Systems*, 28:836–866, 2014.

[Frantz *et al.*, 2013] C. Frantz, M. K. Purvis, M. Nowostawski, and B. T. R. Savarimuthu. nADICO: A nested grammar of institutions. In G. Boella, E. Elkind, B. T. R. Savarimuthu, F. Dignum, and M. K. Purvis, editors, *PRIMA 2013: Principles and Practice of Multi-Agent Systems*, volume 8291 of *Lecture Notes in Artificial Intelligence*, pages 429–436, Berlin, 2013. Springer.

[Frantz et al., 2014a] C. Frantz, M. K. Purvis, M. Nowostawski, and B. T. R. Savarimuthu. Modelling institutions using dynamic deontics. In T. Balke, F. Dignum, M. B. van Riemsdijk, and A. K. Chopra, editors, *Coordination, Organizations, Institutions and Norms in Agent Systems IX*, volume 8386 of *Lecture Notes in Artificial Intelligence*, pages 211–233, Berlin, 2014. Springer.

[Frantz et al., 2014b] C. Frantz, M. K. Purvis, B. T. R. Savarimuthu, and M. Nowostawski. Analysing the dynamics of norm evolution using interval type-2 fuzzy sets. In *WI-IAT '14 Proceedings of the 2014 IEEE/WIC/ACM International Joint Conferences on Web Intelligence (WI) and Intelligent Agent Technologies (IAT)*, volume 3, pages 230–237, 2014.

[Frantz et al., 2014c] C. Frantz, M. K. Purvis, B. T. R. Savarimuthu, and M. Nowostawski. Modelling dynamic normative understanding in agent societies. In H. K. Dam, J. Pitt, Y. Xu, G. Governatori, and T. Ito, editors, *Principles and Practice of Multi-Agent Systems - 17th International Conference, PRIMA 2014*, volume 8861 of *Lecture Notes in Artificial Intelligence*, pages 294–310, Berlin, 2014. Springer.

[Frantz et al., 2015] C. K. Frantz, M. K. Purvis, B. T. R. Savarimuthu, and M. Nowostawski. Modelling dynamic normative understanding in agent societies. *Scalable Computing: Practice and Experience*, 16(4):355–378, 2015.

[Frantz et al., 2016] C. K. Frantz, B. T. R. Savarimuthu, M. K. Purvis, and M. Nowostawski. Generalising social structure using interval type-2 fuzzy sets. In M. Baldoni, A. K. Chopra, T. C. Son, K. Hirayama, and P. Torroni, editors, *PRIMA 2016: Principles and Practice of Multi-Agent Systems: 19th International Conference, Phuket, Thailand, August 22-26, 2016, Proceedings*, pages 344–354. Springer International Publishing, Cham, 2016.

[Frantz, 2015] C. K. Frantz. *Agent-Based Institutional Modelling: Novel Techniques for Deriving Structure from Behaviour*. PhD thesis, University of Otago, Dunedin, New Zealand, 2015. Available under: http://hdl.handle.net/10523/5906.

[Galan and Izquierdo, 2005] J. M. Galan and L. R. Izquierdo. Appearances can be deceiving: lessons learned re-implementing Axelrod's 'evolutionary approach to norms'. *Journal of Artificial Societies and Social Simulation*, 8(3):2, 2005.

[Goette et al., 2006] L. Goette, D. Huffman, and S. Meier. The impact of group membership on cooperation and norm enforcement: Evidence using random assignment to real social groups. *American Economic Review*, 96(2):212–216, May 2006.

[Governatori and Rotolo, 2010] G. Governatori and A. Rotolo. Changing legal systems: legal abrogations and annulments in defeasible logic. *Logic Journal of IGPL*, 18(1):157–194, 2010.

[Governatori et al., 2013] G. Governatori, A. Rotolo, F. Olivieri, and S. Scannapieco. Legal contractions: A logical analysis. In E. Francesconi and B. Verheij, editors, *ICAIL*, pages 63–72. ACM, 2013.

[Greenwald et al., 2002] A. G. Greenwald, M. R. Banaji, L. A. Rudman, S. D. Farnham, B. A. Nosek, and D. S. Mellott. A unified theory of implicit attitudes, stereotypes, self-esteem, and self-concept. *Psychological Review*, 109(1):3–25, 2002.

[Griffiths and Luck, 2010] N. Griffiths and M. Luck. Norm diversity and emergence in tag-based cooperation. In M. D. Vos, N. Fornara, J. V. Pitt, and G. A. Vouros, editors, *Coordination, Organizations, Institutions, and Norms in Agent Systems VI - COIN 2010 International Workshops, COIN@AAMAS 2010, Toronto, Canada, May 2010, COIN@MALLOW 2010, Lyon, France, Au-*

gust 2010, Revised Selected Papers, volume 6541 of Lecture Notes in Computer Science, pages 230–249. Springer, 2010.

[Grossi et al., 2008] D. Grossi, J.-J. Meyer, and F. Dignum. The many faces of counts-as: A formal analysis of constitutive-rules. J. of Applied Logic, 6(2):192–217, 2008.

[Hales, 2002] D. Hales. Group reputation supports beneficent norms. Journal of Artificial Societies and Social Simulation, 5(4):4, 2002.

[Hansen et al., 2007] J. Hansen, G. Pigozzi, and L. van der Torre. Ten philosophical problems in deontic logic. In G. Boella, L. van der Torre, and H. Verhagen, editors, Normative Multi-Agent Systems. Dagstuhl Seminar Proc. 07122, 2007.

[Hansson, 1993] S. Hansson. Reversing the Levi identity. Journal of Philosophical Logic, 22:637–669, 1993.

[Henderson, 2005] D. Henderson. Norms, invariance, and explanatory relevance. Philosophy of the Social Sciences, 35(3):324–338, 2005.

[Hoffmann, 2003] M. Hoffmann. Entrepreneurs and norm dynamics: An agent-based model of the norm life cycle. Technical report, Department of Political Science and International Relations, University of Delaware, Newark (DE), 2003.

[Hoffmann, 2005] M. Hoffmann. Self-organized criticality and norm avalanches. In Proceedings of the Symposium on Normative Multi-Agent Systems, Hatfield (UK), 2005. AISB.

[Hogg, 2001] M. A. Hogg. A social identity theory of leadership. Personality and Social Psychology Review, 5(3):184–200, 2001.

[Hollander and Wu, 2011a] C. Hollander and A. Wu. Using the process of norm emergence to model consensus formation. In Fifth IEEE International Conference on Self-Adaptive and Self-Organizing Systems (SASO), pages 148–157, Oct 2011.

[Hollander and Wu, 2011b] C. D. Hollander and A. S. Wu. The current state of normative agent-based systems. Journal of Artificial Societies and Social Simulation, 14(2):6, 2011.

[Horne, 2001] C. Horne. Sociological perspectives on the emergence of norms. In M. Hechter and K. Opp, editors, Social Norms, pages 3–34. Russell Sage Foundation, New York (NY), 2001.

[Horne, 2007] C. Horne. Explaining norm enforcement. Rationality and Society, 19(2):139–170, 2007.

[Huang et al., 2016] X. Huang, J. Ruan, Q. Chen, and K. Su. Normative multiagent systems: A dynamic generalization. In Proceedings of the Twenty-Fifth International Joint Conference on Artificial Intelligence, IJCAI'16, pages 1123–1129. AAAI Press, 2016.

[Huber, 1999] M. J. Huber. JAM: A BDI-theoretic mobile agent architecture. In O. Etzioni, J. P. Müller, and J. M. Bradshaw, editors, Proceedings of the Third International Conference on Autonomous Agents (Agents '99), pages 236–243, Seattle (WA), 1999.

[Kahneman and Tversky, 1972] D. Kahneman and A. Tversky. Subjective probability: A judgment of representativeness. Cognitive Psychology, 3(3):430 – 454, 1972.

[Kittock, 1995] J. Kittock. Emergent conventions and the structure of multi-agent systems. In Lectures in Complex systems: the proceedings of the 1993 Complex systems summer school, Santa Fe Institute Studies in the Sciences of Complexity Lecture Volume VI, Santa Fe Institute, pages 507–521. Addison-Wesley, 1995.

[Lewis, 1969] D. K. Lewis. *Convention: A Philosophical Study*. Harvard University Press, Cambridge (MA), 1969.

[López y López and Luck, 2004] F. López y López and M. Luck. Towards a model of the dynamics of normative multi-agent systems. In G. Lindemann, D. Moldt, and M. Paolucci, editors, *RASTA 2002*, volume 2934 of *Lecture Notes in Artificial Intelligence*, pages 175–194, Heidelberg, 2004. Springer.

[López y López and Márquez, 2004] F. López y López and A. A. Márquez. An architecture for autonomous normative agents. In *Proceedings of the Fifth Mexican International Conference in Computer Science - ENC*, pages 96–103, Los Alamitos, CA, USA, 2004. IEEE Computer Society.

[López y López et al., 2002] F. López y López, M. Luck, and M. d'Inverno. Constraining autonomy through norms. In *Proceedings of the First International Joint Conference on Autonomous Agents and Multiagent Systems AAMAS*, pages 674–681, New York, NY, USA, 2002. ACM.

[López y López et al., 2006] F. López y López, M. Luck, and M. d'Inverno. A normative framework for agent-based systems. *Computational & Mathematical Organization Theory*, 12(2):227–250, 2006.

[López y López et al., 2007] F. López y López, M. Luck, and M. d'Inverno. A normative framework for agent-based systems. In G. Boella, L. van der Torre, and H. Verhagen, editors, *Normative Multi-agent Systems, Dagstuhl Seminar Proceedings 07122*, Dagstuhl Seminar Proceedings. Internationales Begegnungs- und Forschungszentrum für Informatik (IBFI), Schloss Dagstuhl, Germany, 2007.

[López y López, 2003] F. López y López. *Social Powers and Norms: Impact on Agent Behaviour*. PhD thesis, Department of Electronics and Computer Science, University of Southampton, United Kingdom, 2003.

[Mahmoud et al., 2012] S. Mahmoud, N. Griffiths, J. Keppens, and M. Luck. Efficient norm emergence through experiential dynamic punishment. In *ECAI'12*, pages 576–581, 2012.

[Mahmoud et al., 2014a] M. A. Mahmoud, M. S. Ahmad, M. Z. M. Yusoff, and A. Mustapha. Norms assimilation in heterogeneous agent community. In H. K. Dam, J. Pitt, Y. Xu, G. Governatori, and T. Ito, editors, *Principles and Practice of Multi-Agent Systems - 17th International Conference, PRIMA 2014*, volume 8861 of *Lecture Notes in Artificial Intelligence*, pages 311–318, Berlin, 2014. Springer.

[Mahmoud et al., 2014b] M. A. Mahmoud, M. S. Ahmad, M. Z. M. Yusoff, and A. Mustapha. A review of norms and normative multiagent systems. *The Scientific World Journal*, 2014:23 pages, 2014. Article ID 684587.

[Mahmoud et al., 2015] S. Mahmoud, N. Griffiths, J. Keppens, A. Taweel, T. J. Bench-Capon, and M. Luck. Establishing norms with metanorms in distributed computational systems. *Artif. Intell. Law*, 23(4):367–407, December 2015.

[Makinson and van der Torre, 2000] D. Makinson and L. van der Torre. Input/output logics. *Journal of Philosophical Logic*, 29:383–408, 2000.

[Makinson and van der Torre, 2003] D. Makinson and L. van der Torre. What is input/output logic. In B. Löwe, W. Malzkom, and T. Räsch, editors, *Foundations of the Formal Sciences II : Applications of Mathematical Logic in Philosophy and Linguistics (Papers of a conference held in Bonn, November 10-13, 2000)*, Trends in Logic, vol. 17, pages 163–174, Dordrecht, 2003. Kluwer.

Reprinted in this volume.

[Manna and Waldinger, 1980] Z. Manna and R. Waldinger. A deductive approach to program synthesis. *ACM Transactions on Programming Languages and Systems*, 2(1):90–121, 1980.

[Maranhão, 2001] J. Maranhão. Refinement. A tool to deal with inconsistencies. In *Proceedings of the 8th ICAIL*, pages 52–59, 2001.

[Maranhão, 2017] J. Maranhão. A logical architecture for dynamic legal interpretation. In *Proceedings of 16th ICAIL*, pages 129–139, 2017.

[Meneguzzi and Luck, 2009] F. Meneguzzi and M. Luck. Norm-based behaviour modification in bdi agents. In *Proceedings of The 8th International Conference on Autonomous Agents and Multiagent Systems - Volume 1*, AAMAS '09, pages 177–184, Richland, SC, 2009. International Foundation for Autonomous Agents and Multiagent Systems.

[Mihaylov *et al.*, 2014] M. Mihaylov, K. Tuyls, and A. Nowé. A decentralized approach for convention emergence in multi-agent systems. *Autonomous Agents and Multi-Agent Systems*, 28(5):749–778, 2014.

[Morales *et al.*, 2013] J. Morales, M. López-Sánchez, J. A. Rodríguez-Aguilar, M. Wooldridge, and W. Vasconcelos. Automated synthesis of normative systems. In *Proceedings of the 2013 International Conference on Autonomous Agents and Multi-Agent Systems*, AAMAS '13, pages 483–490, 2013.

[Morales *et al.*, 2014] J. Morales, M. López-Sánchez, J. A. Rodríguez-Aguilar, M. Wooldridge, and W. Vasconcelos. Minimality and simplicity in the on-line automated synthesis of normative systems. In *Proceedings of the 2014 International Conference on Autonomous Agents and Multi-Agent Systems*, AAMAS '14, pages 109–116, 2014.

[Morales *et al.*, 2015a] J. Morales, M. López-Sánchez, J. A. Rodriguez-Aguilar, W. Vasconcelos, and M. Wooldridge. Online automated synthesis of compact normative systems. *ACM Trans. Auton. Adapt. Syst.*, 10(1):2:1–2:33, March 2015.

[Morales *et al.*, 2015b] J. Morales, M. López-Sánchez, J. A. Rodríguez-Aguilar, M. Wooldridge, and W. Vasconcelos. Synthesising liberal normative systems. In *Proceedings of the 2015 International Conference on Autonomous Agents and Multiagent Systems*, AAMAS '15, pages 433–441, Richland, SC, 2015. International Foundation for Autonomous Agents and Multiagent Systems.

[Morales *et al.*, 2015c] J. Morales, I. Mendizabal, D. Sanchez-Pinsach, M. López-Sánchez, and J. A. Rodríguez-Aguilar. Using IRON to build frictionless on-line communities. *AI Commun.*, 28(1):55–71, 2015.

[Mukherjee *et al.*, 2007] P. Mukherjee, S. Sen, and S. Airiau. Emergence of norms with biased interaction in heterogeneous agent societies. In *Proceedings of the 2007 IEEE/WIC/ACM International Conferences on Web Intelligence and Intelligent Agent Technology*, pages 512–515, 2007.

[Mukherjee *et al.*, 2008] P. Mukherjee, S. Sen, and S. Airiau. Norm emergence under constrained interactions in diverse societies. In *Proceedings of the 7th International Joint Conference on Autonomous Agents and Multiagent Systems - Volume 2*, AAMAS '08, pages 779–786, Richland, SC, 2008. International Foundation for Autonomous Agents and Multiagent Systems.

[Nadelman, 1990] E. Nadelman. Global prohibition regimes: The evolution of norms in international society. *International Organization*, 44(4):479–526, 1990.

[Nakamaru and Levin, 2004] M. Nakamaru and S. A. Levin. Spread of two linked social norms on complex interaction networks. *Journal of Theoretical Biology*, 230(1):57–64, 2004.

[Onn and Tennenholtz, 1997] S. Onn and M. Tennenholtz. Determination of social laws for multi-agent mobilization. *Artificial Intelligence*, 95(1):155–167, 1997.

[Ossowski, 2013] S. Ossowski. *Agreement Technologies*. Springer, Dordrecht (NL), 2013.

[Perreau de Pinninck *et al.*, 2008] A. Perreau de Pinninck, C. Sierra, and M. Schorlemmer. Distributed norm enforcement via ostracism. In J. Sichman, J. Padget, S. Ossowski, and P. Noriega, editors, *Coordination, Organizations, Institutions, and Norms in Agent Systems III*, volume 4870 of *Lecture Notes in Computer Science*, pages 301–315. Springer, Berlin, 2008.

[Perreau de Pinninck *et al.*, 2010] A. Perreau de Pinninck, C. Sierra, and M. Schorlemmer. A multiagent network for peer norm enforcement. *Autonomous Agents and Multi-Agent Systems*, 21(3):397–424, 2010.

[Pigozzi and van der Torre, 2017] G. Pigozzi and L. van der Torre. Multiagent deontic logic and its challenges from a normative systems perspective. *The IfCoLog Journal of Logics and their Applications*, 4(9), 2017.

[Pucella and Weissman, 2004] R. Pucella and V. Weissman. Reasoning about dynamic policies. In I. Walukiewicz, editor, *FOSSACS 2004. LNCS, vol. 2987*, pages 453–467. Springer, Heidelberg, 2004.

[Pujol *et al.*, 2005] J. M. Pujol, J. Delgado, R. Sangüesa, and A. Flache. The role of clustering on the emergence of efficient social conventions. In *Proceedings of the 19th International Joint Conference on Artificial Intelligence*, IJCAI'05, pages 965–970, San Francisco, CA, USA, 2005. Morgan Kaufmann Publishers Inc.

[Rao and Georgeff, 1995] A. S. Rao and M. P. Georgeff. BDI agents: From theory to practice. *Proceedings of the First International Conference on Multi-Agent Systems (ICMAS-95)*, pages 312–319, 1995.

[Risse and Sikkink, 1999] S. R. Risse, Thomas and K. Sikkink. *The Power of Human Rights: International Norms and Domestic Change*. Cambridge University Press, Cambridge, 1999.

[Riveret *et al.*, 2012] R. Riveret, A. Rotolo, and G. Sartor. Probabilistic rule-based argumentation for norm-governed behaviour. *Artificial Intelligence*, 20(4):383–420, 2012.

[Riveret *et al.*, 2013] R. Riveret, G. Contissa, D. Busquets, A. Rotolo, J. Pitt, and G. Sartor. Vicarious reinforcement and ex ante law enforcement: A study in norm-governed learning agents. In *Proceedings of the Fourteenth International Conference on Artificial Intelligence and Law*, ICAIL '13, pages 222–226, New York, NY, USA, 2013. ACM.

[Riveret *et al.*, 2014] R. Riveret, A. Artikis, D. Busquets, and J. Pitt. Self-governance by transfiguration: From learning to prescriptions. In F. Cariani, D. Grossi, J. Meheus, and X. Parent, editors, *Deontic Logic and Normative Systems*, volume 8554 of *Lecture Notes in Computer Science*, pages 177–191. Springer, Springer, 2014.

[Russell *et al.*, 2015] S. Russell, D. Dewey, and M. Tegmark. Research priorities for robust and beneficial artificial intelligence. *AI Magazine*, 36(4):105–114, 2015.

[Saam and Harrer, 1999] N. J. Saam and A. Harrer. Simulating norms, social inequality, and functional change in artificial societies. *Journal of Artificial Societies and Social Simulation*, 2(1):2, 1999.

[Salazar *et al.*, 2010] N. Salazar, J. A. Rodriguez-Aguilar, and J. L. Arcos. Robust coordination in large convention spaces. *AI Communications*, 23(4):357–372, 2010.

[Savarimuthu *et al.*, 2011] B. Savarimuthu, R. Arulanandam, and M. Purvis. Aspects of active norm learning and the effect of lying on norm emergence in agent societies. In D. Kinny, J.-j. Hsu, G. Governatori, and A. Ghose, editors, *Agents in Principle, Agents in Practice*, volume 7047 of *Lecture Notes in Computer Science*, pages 36–50. Springer, Berlin, 2011.

[Savarimuthu and Cranefield, 2009] B. T. R. Savarimuthu and S. Cranefield. A categorization of simulation works on norms. In G. Boella, G. Pigozzi, and L. van der Torre, editors, *Normative Multi-agent Systems, Dagstuhl Seminar Proceedings 09121*, pages 39–58, Internationales Begegnungs- und Forschungszentrum für Informatik (IBFI), Schloss Dagstuhl, Germany, 2009.

[Savarimuthu and Cranefield, 2011] B. T. R. Savarimuthu and S. Cranefield. Norm creation, spreading and emergence: A survey of simulation models of norms in multi-agent systems. *Multiagent and Grid Systems*, 7(1):21–54, January 2011.

[Savarimuthu *et al.*, 2007] B. T. R. Savarimuthu, S. Cranefield, M. Purvis, and M. Purvis. Norm emergence in agent societies formed by dynamically changing networks. In *2007 IEEE/WIC/ACM International Conference on Intelligent Agent Technology*, pages 464–470, 2007.

[Savarimuthu *et al.*, 2008a] B. T. R. Savarimuthu, S. Cranefield, M. Purvis, and M. Purvis. Role model based mechanism for norm emergence in artificial agent societies. In J. Sichman, J. Padget, S. Ossowski, and P. Noriega, editors, *Coordination, Organizations, Institutions, and Norms in Agent Systems III*, volume 4870 of *Lecture Notes in Computer Science*, pages 203–217. Springer, Berlin, 2008.

[Savarimuthu *et al.*, 2008b] B. T. R. Savarimuthu, M. A. Purvis, and M. K. Purvis. Social norm emergence in virtual agent societies. In *Proceedings of the 7th International Conference on Autonomous Agents and Multi-Agent Systems*, 2008.

[Savarimuthu *et al.*, 2009] B. T. R. Savarimuthu, S. Cranefield, M. A. Purvis, and M. K. Purvis. Norm emergence in agent societies formed by dynamically changing networks. *Web Intelligence and Agent Systems*, 7(3):223–232, 2009.

[Savarimuthu *et al.*, 2010a] B. T. R. Savarimuthu, S. Cranefield, M. A. Purvis, and M. K. Purvis. A data mining approach to identify obligation norms in agent societies. In *Proceedings of the International Workshop on Agents and Data Mining Interaction (ADMI@AAMAS 2010), Toronto, Canada*, pages 54–69, May 2010.

[Savarimuthu *et al.*, 2010b] B. T. R. Savarimuthu, S. Cranefield, M. A. Purvis, and M. K. Purvis. Obligation norm identification in agent societies. *Journal of Artificial Societies and Social Simulation*, 13(4):3, 2010.

[Savarimuthu *et al.*, 2013a] B. T. R. Savarimuthu, S. Cranefield, M. A. Purvis, and M. K. Purvis. Identifying prohibition norms in agent societies. *Artificial Intelligence and Law*, 21:1–46, 2013.

[Savarimuthu *et al.*, 2013b] B. T. R. Savarimuthu, J. Padget, and M. A. Purvis. Social norm recommendation for virtual agent societies. In G. Boella, E. Elkind, B. T. R. Savarimuthu, F. Dignum, and M. K. Purvis, editors, *PRIMA 2013: Principles and Practice of Multi-Agent Systems*, volume

8291 of *Lecture Notes in Computer Science*, pages 308–323. Springer, Berlin, 2013.

[Savarimuthu, 2011] B. T. R. Savarimuthu. *Mechanisms for Norm Emergence and Norm Identification in Multi-Agent Societies*. PhD thesis, University of Otago, Dunedin, New Zealand, 2011.

[Scheve *et al.*, 2006] C. v. Scheve, D. Moldt, J. Fix, and R. v. Luede. My agents love to conform: Norms and emotion in the micro-macro link. *Computational & Mathematical Organization Theory*, 12(2):81–100, 2006.

[Schneider and Teske, 1992] M. Schneider and P. Teske. Toward a theory of the political entrepreneur: Evidence from local government. *American Political Science Review*, 86(3):737–747, 1992.

[Sen and Airiau, 2007] S. Sen and S. Airiau. Emergence of norms through social learning. In M. Veloso, editor, *Proceedings of the 20th International Joint Conference on Artifical Intelligence*, IJCAI'07, pages 1507–1512, San Francisco (CA), 2007. Morgan Kaufmann Publishers Inc.

[Sen and Sen, 2010] O. Sen and S. Sen. Effects of social network topology and options on norm emergence. In J. Padget, A. Artikis, W. Vasconcelos, K. Stathis, V. da Silva, E. Matson, and A. Polleres, editors, *Coordination, Organizations, Institutions and Norms in Agent Systems V*, volume 6069 of *Lecture Notes in Computer Science*, pages 211–222. Springer, Berlin, 2010.

[Shoham and Tennenholtz, 1992a] J. Shoham and M. Tennenholtz. Emergent conventions in multi-agent systems: Initial experimental results and observations. In *Proceedings of the Third International Conference on the Principles of Knowledge Representation and Reasoning KR*, pages 225–231, San Mateo, CA, USA, 1992. Morgan Kaufmann.

[Shoham and Tennenholtz, 1992b] Y. Shoham and M. Tennenholtz. On the synthesis of useful social laws for artificial agent societies. In *Proceedings of the 10th National Conference on Artificial Intelligence (AAAI '92)*, pages 276–281, San Jose (CA), July 1992.

[Shoham and Tennenholtz, 1995] Y. Shoham and M. Tennenholtz. On social laws for artificial agent societies: off-line design. *Artificial Intelligence*, 73(1û2):231 – 252, 1995. Computational Research on Interaction and Agency, Part 2.

[Shoham and Tennenholtz, 1997] Y. Shoham and M. Tennenholtz. On the emergence of social conventions: modeling, analysis, and simulations. *Artificial Intelligence*, 94(1–2):139–166, July 1997.

[Simon, 1955] H. A. Simon. A Behavioral Model of Rational Choice. *The Quarterly Journal of Economics*, 69(1):99–118, 1955.

[Sims and Brinkmann, 2003] R. R. Sims and J. Brinkmann. Enron ethics (or: Culture matters more than codes). *Journal of Business Ethics*, 45(3):243–256, Jul 2003.

[Singh, 2014] M. P. Singh. Norms as a basis for governing sociotechnical systems. *ACM Trans. Intell. Syst. Technol.*, 5(1):21:1–21:23, January 2014.

[Staller and Petta, 2001] A. Staller and P. Petta. Introducing emotions into the computational study of social norms: A first evaluation. *Journal of Artificial Societies and Social Simulation*, 4(1):2, 2001.

[Stolpe, 2010] A. Stolpe. Norm-system revision: Theory and application. *Artificial Intelligence and Law*, 18:247–283, 2010.

[Sugawara, 2011] T. Sugawara. Emergence and stability of social conventions in conflict situations. In *Proceedings of the Twenty-Second International Joint Conference on Artificial Intelligence - Volume Volume One*, IJCAI'11, pages 371–378. AAAI Press, 2011.

[Sunstein, 1996] C. R. Sunstein. Social norms and social roles. *Columbia Law Review*, 96(4):903–968, 1996.

[Tinnemeier *et al.*, 2009] N. Tinnemeier, M. Dastani, J.-J. C. Meyer, and L. van der Torre. Programming normative artifacts with declarative obligations and prohibitions. In *Proc. of WI/IATÕ09*. IEEE Computer Society, 2009.

[Tinnemeier *et al.*, 2010] N. Tinnemeier, M. Dastani, and J.-J. Meyer. Programming norm change. In *Proceedings of the 9th International Conference on Autonomous Agents and Multiagent Systems: Volume 1 - Volume 1*, AAMAS '10, pages 957–964, Richland, SC, 2010. International Foundation for Autonomous Agents and Multiagent Systems.

[Urbano *et al.*, 2009] P. Urbano, J. Balsa, L. Antunes, and L. Moniz. Force versus majority: A comparison in convention emergence efficiency. In J. Hübner, E. Matson, O. Boissier, and V. Dignum, editors, *Coordination, Organizations, Institutions and Norms in Agent Systems IV*, volume 5428 of *Lecture Notes in Computer Science*, pages 48–63. Springer, Berlin, 2009.

[Valk, 1998] R. Valk. Petri nets as token objects. an introduction to elementary. In *Proceedings of Application and Theory of Petri Nets*, pages 1–25, Berlin, 1998. Springer.

[van Ditmarsch and van der Hoek, 2007] H. van Ditmarsch and K. B. van der Hoek, W. *Dynamic Epistemic Logic*. Synthese Library Series, vol. 337, Springer, Heidelberg, 2007.

[Verhagen, 2000] H. J. Verhagen. *Norm Autonomous Agents*. PhD thesis, The Royal Institute of Technology and Stockholm University, Stockholm, Sweden, 2000.

[Verhagen, 2001] H. Verhagen. Simulation of the learning of norms. *Social Science Computer Review*, 19(3):296–306, 2001.

[Villatoro *et al.*, 2009] D. Villatoro, S. Sen, and J. Sabater-Mir. Topology and memory effect on convention emergence. In *Proceedings of the 2009 IEEE/WIC/ACM International Joint Conference on Web Intelligence and Intelligent Agent Technology - Volume 02*, WI-IAT '09, pages 233–240, 2009.

[Villatoro *et al.*, 2011a] D. Villatoro, G. Andrighetto, J. Sabater-Mir, and R. Conte. Dynamic sanctioning for robust and cost-efficient norm compliance. In T. Walsh, editor, *Proceedings of the 22nd International Joint Conference on Artificial Intelligence*, volume 1 of *IJCAI'11*, pages 414–419. AAAI Press, 2011.

[Villatoro *et al.*, 2011b] D. Villatoro, J. Sabater-Mir, and S. Sen. Social instruments for robust convention emergence. In T. Walsh, editor, *Proceedings of the 22nd International Joint Conference on Artificial Intelligence*, volume 1 of *IJCAI'11*, pages 420–425. AAAI Press, 2011.

[Villatoro *et al.*, 2013] D. Villatoro, J. Sabater-Mir, and S. Sen. Robust convention emergence in social networks through self-reinforcing structures dissolution. *ACM Trans. Auton. Adapt. Syst.*, 8(1):2:1–2:21, April 2013.

[Walker and Wooldridge, 1995] A. Walker and M. Wooldridge. Understanding the emergence of conventions in multi-agent systems. In *Proceedings of the first international conference on multi-agent systems (ICMAS)*, pages 384–389, Menlo Park (CA), 1995. AAAI Press.

[Watkins and Dayan, 1992] C. Watkins and P. Dayan. Technical note: Q-learning. *Machine Learning*, 8(3-4):279–292, 1992.

[Watts and Strogatz, 1998] D. J. Watts and S. H. Strogatz. Collective dynamics of 'small-world' networks. *Nature*, 393(6684):440–442, June 1998.

[Young, 1990] O. Young. Political leadership and regime formation: On the development of institutions in international society. *International Organization*, 45(3):281–308, 1990.

[Younger, 2004] S. Younger. Reciprocity, normative reputation, and the development of mutual obligation in gift-giving societies. *Journal of Artificial Societies and Social Simulation*, 7(1):5, 2004.

[Yu *et al.*, 2010] C.-H. Yu, J. Werfel, and R. Nagpal. Collective decision-making in multi-agent systems by implicit leadership. In *Proceedings of the 9th International Conference on Autonomous Agents and Multiagent Systems: Volume 3 - Volume 3*, AAMAS '10, pages 1189–1196, Richland, SC, 2010. International Foundation for Autonomous Agents and Multiagent Systems.

[Yu *et al.*, 2013] C. Yu, M. Zhang, F. Ren, and X. Luo. Emergence of social norms through collective learning in networked agent societies. In *Proceedings of the 2013 International Conference on Autonomous Agents and Multi-agent Systems*, AAMAS '13, pages 475–482, 2013.

[Yu *et al.*, 2015] C. Yu, H. Lv, F. Ren, H. Bao, and J. Hao. Hierarchical learning for emergence of social norms in networked multiagent systems. In *AI 2015: Advances in Artificial Intelligence: 28th Australasian Joint Conference, Canberra, ACT, Australia, November 30 – December 4, 2015, Proceedings*, pages 630–643, Cham, 2015. Springer International Publishing.

[Zhang and Leezer, 2009] Y. Zhang and J. Leezer. Emergence of social norms in complex networks. In *International Conference on Computational Science and Engineering*, pages 549–555, Vancouver, 2009.

Received 5 September 2016

Modeling Organizations and Institutions in MAS

Nicoletta Fornara
*Università della Svizzera italiana,
via G. Buffi 13, 6900 Lugano, Switzerland
nicoletta.fornara@usi.ch*

Tina Balke-Visser
*University of Surrey
Guildford, UK
tina.balke@gmail.com*

Abstract

Institutions and *Organizations* are two concepts within the MAS community that are commonly referred to when the question arises on how to ensure that an (open) MAS exhibits some desired properties, while the agents interacting in that MAS have some degree of autonomy at the same time. This chapter gives a brief introduction to the two concepts and its related ideas. It outlines research done in the area of normative MAS and gives pointers on current challenges for modeling institutions and organizations.

1 Introduction

In Multi-Agents systems (MAS), software agents that enjoy some degree of autonomy interact [70]. As a consequence, similar to human societies, the problem arises on how to ensure that the MAS exhibits some desired global property, without compromising the agent's autonomy at the same time [41, p. 2]. Leaning on existing works such as for example in sociology, psychology and organizational theory, in recent years MAS researchers have been starting to incorporate and model concepts such organizations and institutions in computational systems, as demonstrated by several publications on the topic in the AAMAS conference series[1], the COIN workshop series[2], and the Normative Multi-Agent Systems seminars[3].

---

[1]http://www.ifaamas.org/proceedings.html
[2]http://www.pcs.usp.br/~coin
[3]http://icr.uni.lu/normas/history.html

In this Chapter, we will provide an introduction to the concepts of institutions and organizations and their modeling. This chapter is not aiming to be an in-depth literature review and it will not give details on all aspects of modeling institutions and organizations in MAS, but it rather aims to point the interest reader to topics and areas of interest and give him or her starting points for further studies.

Wanting to model "institutions" and "organizations" a first step is to understand what the two words means and what concept they refer to. As simple as this sounds, this task is not an easy one as (i) not only are the two concepts interlinked - they are both broadly speaking, coordinate means [41] - but (ii) in the agents community, the words are often used as synonymous. One of the reasons for the latter is that different research communities started to use the terms differently, sometimes borrowing concepts from other disciplines. Researcher wanting to publish/work in the respective communities – in order to pass review processes for their papers – had to use the communities jargon, i.e. use the terms the community was using. This resulted in situations where researchers used different words for one and the same idea they described, in order to publish in the different communities they were working in.

Taking a step back, a popular source, often cited by the agent community when it comes to the definition of the terms "institutions" and "organizations", is North [54, p. 4f]:

> "A crucial distinction in this study is made between institutions and organizations. [...] Conceptually what must be clearly differentiated are the rules from the players. The purpose of the rules is to define the way the game is played. But the objective of the team within that set of rules is to win the game – by a combination of skills, strategy and coordination; by fair means and sometimes by foul means. Modeling the strategies and the skills of the team as it develops is a separate process from modeling the creation, evolution, and consequences of the rules."

The distinction indicated by North is the idea that organizations are agents like households, firms and states that have preferences and objectives, whereas institution are formal and informal societal constraints such as laws, conventions, constitutions, habits and rules, which reduce the total scarce resources available [48]. Broadly speaking, both institutions as well as organizations are means of coordination and provide some form of structure, but whereas institutions focus on the structure of the rules and norms, organizational structure of a MAS concerns the agents, their roles and their relationships by which the overall behavior of the MAS is defined [41].

Based on this abstract distinction, this chapter tries to give an overview of both, modeling organizations as well as institutions (and the differences between them). For this purpose, we start by looking at the modeling of institutions first, by discussing regulative and constitutive norms, as well as agents communication languages as means to communicate and thereby share norms. Afterwards the focus of the chapter will turn to organizations with the topics of modeling agents (and their roles) as well as their relations to one another in terms of organizational structures are being addressed.

2   Survey on Modeling Institutions in MAS

The formalization, realization, and management of open distributed interaction systems where autonomous software agents (operating on behalf of one or more human users) may interact for exchanging resources or providing services is widely recognized to be an important research problem that is becoming more and more relevant with the massive development of distributed social network systems on the Internet.

One approach, which may be followed for the formalization of those systems, consists in modeling them as a set of *artificial institutions* (AI). Human social institutions[60], like for example the institution of *marriage* or *family*, the institution of *money*, or the institution of *education*, have been used, in Multiagent Systems (MAS) research, as a source of inspiration for the definition of the various abstract concepts and software components required for the concrete realization of artificial/electronic institutions. Artificial institutions are fundamental because:

> "Their main purpose is to **enable** and **regulate** the interaction among autonomous agents in order to achieve some collective endeavour" [30, p. 278].

In *open distributed systems* or *socio-technical systems* [12], which support the interaction of various components (autonomous software called agents or humans) with different, often competitive, goals, there is the need to *enable* and to *regulate* such interactions. This with the objective to keep the evolution of those systems with-in certain boundaries, and allows the system itself to reach certain social goals. There is also the need to create in the interacting agents an *expectation* on the reasonable future evolution of the interaction, this in order to enable the agents to coherently plan their actions for reaching their own goals. This can be done by formalizing an open distributed system using multiple, sometime interconnected, artificial institutions, and by modelling and realizing certain software components for their management.

In order to create *open spaces* where interactions among autonomous agents may happen and where those interactions may be constrained without being always regimented, it is necessary to analyze and formally specify the various interconnected *static and dynamic components* that enable and regulate those interactions in real life. It is therefore necessary to:

1. Formally define the application-independent concepts that are relevant for the definition of agents' institutions, for example the notion of *norm* or *regulation*, *institutional power*, and *constitutive rule*;

2. Specify the software components required for the management of concrete the objects created from the abstract concepts, like for example a *norm monitoring* component, or a component for computing the state of the interaction among agents on the basis of the concrete institutional powers assigned to the agents in a system;

3. Formally define the conceptual and logical model of the application dependent knowledge/data used by those software components.

The choice of the formalism to adopt for the specification and implementation of the various software components and for the specification of the abstract concepts and the concrete definition of instances of those concepts is an important aspect in the definition of a model for artificial institutions.

In MAS literature, as discussed below, there are various proposals for the formalization of artificial institutions, which are also called electronic institutions or agent institutions. The conceptual model of the fundamental concepts required for the formal specification of those institutions is usually placed side by side with the specification of an institutional framework required for the actual implementation of institutions and of the software components for their management.

A recent and very interesting discussion on the analogies and differences between artificial/electronic institutions (AI) and virtual organizations plus an extensive comparison of three models for the formal specification of institutions are presented in [30][4]. The three discussed and compared models are: (i) the ANTE Framework [49]; (ii) the OCeAN metamodel for the specification of Artificial Institutions (AI) [28, 29] that has been extended into the MANET model for the specification of AI situated in environments [67]; (iii) and the conceptual model and computational architecture for Electronic Institutions (EIs) extended with the EIDE development environment [2, 24].

---

[4]This chapter is part of the book "Agreement Technologies" [55], published as result of the COST Action on Agreement Technologies[5] (2008-2012) which involved many researchers in the field of institutional and normative multiagent systems.

Other interesting frameworks, which are briefly discussed in [30], are: (i) the *OMNI* model [21] which allows the description of MAS-based organizations followed by the *OperettA* framework [1] which supports the implementation of real systems; (ii) the *instAL* normative framework [15, 18] that may be used to specify, verify and reason about norms used to regulate MAS; (iii) and the definition of Norm-Governed Computational Societies [5] followed by the specification of Sustainable Institutions [58] which is influenced by Olstrom definition of institution [56].

Another quite recent and relevant book chapter on regulated MAS, which can be used to create an institutional reality where autonomous agents may interact, is the chapter "Regulated MAS: Social Perspective" [53]. It is part of the book on Normative Multi-Agent Systems that was published as result of NorMAS 2012 Seminar, which took place in March 2012 in Dagstuhl, Germany.

Crucial abstract concepts which are required for the specification of every artificial institution and the software components for their management are discussed in the following sections.

2.1   Regulative rules - Norms

A very important characteristic of the autonomous agents developed by different users that interact on an open network, like for example a peer to peer network in Internet, is that no assumption can be made on their internal design and, like for human beings, it is impossible to assume that they will always fulfill their norms. In particular as discussed in [34] it is not always possible and advantageous to regiment all obligations. Think for example to the obligation to pay for an ordered product when it is received, at least, it is reasonable to sanction irregular behaviors. In MAS research there are numerous proposals to formally and declaratively specify norms or policies (these two terms are very often used as synonymous), they are usually used for expressing *obligations*, *permissions*, or *prohibitions* to perform certain actions when certain specific conditions, related to state of affairs or to specific events, are satisfied. Therefore norms are usually characterized by some attributes: the *type*, for distinguishing between obligations, permissions and prohibitions, the *debtor* or the *addressee* which may be expressed using *roles*, the *activation* and the *expiration* or *deactivation conditions* which describe the events or the state of affairs that activate or deactivate the norm, the *content* that is the prohibited, permitted or obliged action or state of affair, and the sanctions and reward for norms fulfillment or violation.

An attribute that may be useful in the specification of abstract norms at design time, is the *roles* that an agent may play in an artificial institution, for example in a auction the role of participant or auctioneer. When one or more institutions are instantiated for the realization of a concrete open interaction system the various roles defined in those institutions have to be replaced by their concrete counterpart that an agent may play in a concrete realization of an artificial institution, like the participant of the auction number run01. Finally when a norm, related to various agents by using the concrete role attribute, becomes active, various instances of that norm (one for each agent that plays the concrete role specified in the norm) need to be created and managed by the interaction system.

An important software component of every normative system is the *monitor of norms*. That is, a component able to compute the state of the norms on the basis of their activation condition and content and on the basis of the actions and events that happen in the system. This component is required in order to be able to compute or deduce if a given norm is fulfilled or violated and therefore apply sanctions or reward with the goal of enforcing norms fulfillment for all those norms that are not regimented. In order that the monitoring component can use the knowledge about the state of the interaction for computing the fulfillment or violation of norms it necessary to realize a *synchronization component* able to dynamically update the *knowledge base* used for representing the state of the interaction with the observable changes due to events or agents actions. The realization of this components may be challenging, in particular for those events that are nor easily observable by means of sensors or that are not directly connected to the actions of the agents. It is also crucial that the *knowledge base* used for representing the state of the interaction has a conceptual model of the concepts and properties (or relations) that are relevant for the description and regulation of the interaction. Starting from general concepts like time, action, event to the specification of application dependent part like for instance in an auction the notion of offer, owner, and the action of paying and delivering.

In an open normative multiagent system, it is crucial to specify the norms using formal declarative languages (like logics or logic programming languages). This choice has many important advantages, because it makes possible to:

- Represent the norms as data, instead of coding them into the software, with the advantage of making possible to add, remove, or change the norms both when the system is off line, and at run-time, without the need to reprogram some components of the interaction system or the software agents that use the system;

- Develop agents able to reason and plan their actions by taking into considerations the correlations between their goals and external social constrains expressed also in terms of norms, this by reasoning on what norms apply in a given situation, what activities are obligatory, permitted or prohibited, using for example some form of what-if reasoning [68] for deciding whether or not to comply to norms by taking into account norm rewards and sanctions;

- Develop agents able to interact within different systems without the need of being reprogrammed [29];

- Realize an application-independent monitoring component able to keep trace of the state of norms on the basis of the events that happen in the system, on the basis of the agents' actions, and on the basis of the state of the interaction (this mechanism can also be able to react to norm fulfilments or violations);

- Realize mechanisms for checking norm conflicts, understanding when conflicts may arise, and solving or avoiding them, like for example by introducing priority ordering between norms [22, 64].

The choice of the formal language used for the declarative specification of norms is difficult because many aspects have to be taken into account. The most important are: the expressivity of the language, its computational complexity, the fact that the underlying logic is decidable, the diffusion of the language among software practitioners and research communities, its feasibility for fast prototyping, and its adoption as an international standard an crucial aspect for having good interoperability between separately engineered software agents.

Semantic Web Technologies [43] may be the successfully adopted for an efficient and effective representation of norms/policies for open interaction systems running on the Internet. A relevant advantage of this choice is that Semantic Web technologies are increasingly becoming a standard for Internet applications and therefore are supported by many tools: many efficient reasoners (like Fact++, Pellet, Racer Pro, HermiT), tools for ontology editing (like Protégé), and libraries for automatic ontology management (like OWL-API and JENA).

Currently there are some works, in the multiagent community, that adopt Semantic Web Technologies for the formalization and management of norms. One is represented by the works of Fornara's group who has investigated the possibility to use OWL 2 and SWRL rules for the specification of agents' commitments and obligations [35, 29, 51]. In those papers an OWL ontology of obligations. The *activation condition* of an obligation is a class of events that when happen trigger the activation of the obligation. The *content* of an obligation is a class of possible actions that have to be performed within a given deadline. The proposed model of obligations allow also to specify the relation between obligations and time, it is therefore possible to specify deadlines and interval of time. The *monitoring* of those obligations, that is checking if they are fulfilled of violated on the basis of the actions of the agents, can be realized thanks to a specific framework required for managing the elapsing of time and to perform closed-world reasoning on certain classes. A similar ontological formalization of obligations has been also extended for being used in a complete OWL 2 model of artificial institutions (initially called OCeAN and subsequential MANET) instantiated at run-time by dynamically creating in the environment *spaces of interaction* [36]. An updated version of such a model of obligations has been also applied in the field of access control where policies has to be specified for regulating the access and the use of data [52, 51].

Another interesting approach that uses Semantic Web Technologies for norms formalization and management is the OWL-POLAR framework for semantic policy representation and reasoning [64]. This framework investigates the possibility of using OWL ontologies for representing the state of the interaction among agents and SPARQL queries for reasoning on policies activation, for anticipating possible conflicts among policies, and for conflicts avoidance and resolution.

Another relevant proposal where Semantic Web technologies are used for policy specification and management is the KAoS policy management framework [68], which is composed by three layers: (i) the human interface layer where policies, expressing authorizations and obligations, are specified in the form of constrained English sentences; (ii) the policy management layer used for encoding in OWL the policy-related information; (iii) the policy monitoring and enforcing layer used to compile OWL policies to an efficient format usable for monitoring and enforcement.

2.2 Constitutive rules

As clearly presented by John Searle [60] in his book "the construction of social reality", in human interactions are involved brute facts and facts that exist only thanks to an institutional setting, this second type of facts are called *institutional facts*. They exist thanks to the existence of a system of *constitutive rules* collectively created that define and create them. *Constitutive rules* have the form *X counts as Y in context C*, where X can be brute fact, Y is an institutional fact and C is the context where the rule holds. Constitutive rules can be used for mapping brute facts into institutional facts. Like for example mapping the raise of one hand into a bid in an auction. Once an institutional fact has been defined, a constitutive rule can be used for specifying that an institutional fact counts as another institutional fact. Like for example mapping the highest bid in an auction into a commitment to pay a given amount of money to the owner of the product sold in the auction.

In artificial institutions that are the digital counterpart or extension of a human institution, the software designers are in charge of deciding: (i) which facts they want to formalize as brute facts, by defining causal effect rules able to change them;,

(ii) which facts they want to formalize as institutional facts, which exist only thanks to common agreement between the interacting agents or between their designers, agreement that may be specified in a system of constitutive rules.

In AI *institutional actions* are a special type of actions that change institutional attributes [28]. Given that institutional attributes exist thanks to the common agreement of the interacting parties, institutional actions can be performed, if certain conditions hold, by means of suitable public communicative acts: declarations. A very important application independent contextual condition that an agent must satisfy in order to successfully perform an institutional action is to hold the *institutional power* to perform it. This connection between a public *declaration* (X) to perform an institutional action by an agent, the holding of various contextual conditions (C) including also the correct institutional power of the agent who is attempting to perform the institutional action, and the correct happening of the institutional action (Y), can be formalized with a *special constitutive rule*. For example an agent can perform a bid in an auction by performing a declaration of the amount of money that it wants to offer, the declaration counts as a bid if the agent is a regular registered participant in the auction.

Like for norms, it is very common to specify the institutional powers of agents at design time by using a set of *roles*, this allows to abstract from the specific agents that will interact within an institutional framework and requires to develop a module for the correct instantiation of the institutional powers at run time. Moreover, similarly to what discussed for norms, it is crucial to implement a *synchronization component* able to dynamically update the *knowledge base* used for representing the state of the interaction with the observable changes due to events or actions that will trigger the various constitutive rules.

The notion of institutional power has been analysed in [46] and it has been discussed and formalized in Event Calculus in [28]. Another logic formalization of constitutive rules can be found in [7] and in [42]. In [10] a recent interesting analysis has been performed on the similarities and differences of using events and states as brute facts for modelling institutional facts.

2.3 Agent Communication

In MAS literature it is possible to identify two different and complementary approaches for supporting and enabling the direct and indirect interactions among agents. A direct interaction may be realized by means of the definition of an Agent Communication Language (ACL) and realized with the direct message passing between agents. An indirect interaction may be realized by using blackboard systems, in which every agent can put information on a common information space, the blackboard, and any agent can read the information from the blackboard at any moment. The main negative aspect of blackboard systems is that they have a centralized structure that is not well suited for the realization of open interaction systems. In the following subsection we will present in more details ACLs.

2.3.1 Agent Communication Languages

In order to interact in an open environment autonomous agents need to adopt a common language, therefore they need to define a standard Agent Communication Language. The most relevant proposals of standard ACL in MAS literature are based on speech act theory [6, 61], an approach that views language use as a form of action, making it possible to treat communicative acts and other types of action in a uniform way.

The first studies on ACLs follow what we can call a *mentalistic* approach, that is, they defines the meaning of a set of communicative acts having different performative by using agent's mental states, like beliefs, desires (or goals) and intentions. Two well-known ACLs that follow this approach are KQML[27] and FIPA ACL[6], proposed by the Foundation for Intelligent Physical Agent (FIPA). Using agents' mental states may be adequate in cooperative multiagent system, but it is not appropriate for open interaction systems composed by heterogeneous and often competitive agents made by different vendors [66]. In this kind of context it is impossible to trust other agents completely or make assumption about their internal design.

Therefore at the beginning of 2000s, a new approach to the definition of ACLs based on the social, objective consequences, and new obligations of performing a speech act were proposed [17, 32]. In this approach, the semantics of different type of speech acts is expressed using *commitments* directed from one agent to another, and this type of ACL is called commitment-based ACL. In particular in this approach, following the taxonomy of speech acts defined by John Searle [62] (which classifies illocutionary acts into five categories: declarations, assertives, commissives, directives and expressives) the semantics of every type of communicative acts is defined using the notion of social commitment, conditional commitment, and pre-commitment.

Agents' commitments to one another can also be used for expressing the semantics of the messages exchanged in the specification of protocols (where protocols represent the allowed interactions among communicating agents). The use of commitments in the specification of the meaning of the messages exchanged in a given protocol makes the specification more flexible with respect to the traditional approaches, which model protocols as fixed action sequences. Two interesting papers

---

[6]http://www.fipa.org/repository/aclspecs.html

on this topic are: [71] where a slightly different model of commitments is used; and [33] where the model of commitment used for the definition of the semantics of commitment-based ACLs is used for the flexible specifications of interactions protocols like the English Auction protocol. In this work it is also showed how it is possible to combine basic acts (belonging to the taxonomy of speech acts) for defining new type of acts that are frequently used in certain applications. For example it is showed that a proposal, used in a lot of e-commerce applications, can be formalized combining a request and a conditional promise to do something on condition that the request will be accepted, this means that it is not simply a combination of two types of communicative acts, but also their content is strongly related.

The approach of using a fixed set of primitives for expressing the semantics of many different types of communicative acts is criticized in a recently published chapter on Agent Communication Languages [13, p.8]. The reason of the critique is based on the fact that in specific application contexts (like business applications) it is not always enough to use basic communicative acts type but it is necessary to define new primitive communicative acts like for example "quote price" or "stock quote". Therefore, in this chapter the semantics of domain-specific primitives, needed by a MAS, is given using a set of abstractions, among them the most important is the notion of commitment. Given that also in the approach based on a fixed set of primitives it is possible to define new communicative acts combining the existing one whose semantics is based on commitments, the two approaches results to be quite similar.

The initial approaches to the definitions of commitment-based ACLs did not define the semantics of a very important type of speech acts: declarations. Declarations are the particular category of communicative acts whose point is to bring about a change in the institutional reality in virtue of their successful performance. Declarations are fundamental in artificial institutions because, as previously discussed, they are the means for performing institutional actions. Their formalization was initially sketched in [31] and improved in [37] where the definition of an ACL is strictly connected with the definition of various artificial institutions.

Finally it is worth to mention that in 2013 a collection of six manifestos, each of which identifies important concerns and directions in agent communication has been published [11].

2.4 Artificial Institutions situated in Environments

Taking into account the literature on modelling agent environment as a first-class abstraction [69], it is possible to extend the model of artificial institutions and the model of the distributed system where AI are used. A crucial task of the environment, in a MAS model, is to register the events or actions that happen in the system and notify them to the agents registered for the template of such events/actions. The realization of this task can be combined with the idea a designing a MAS using different AIs.

In fact, the specification of an artificial institution (AI) consists in the abstract specification at design time of the concepts introduced so far, for example the formalization of norms and constitutive rules in terms of roles. The advantage of such an abstract specification is that it can be re-used in the specification of different MAS in different applications. Once one or more AIs are designed, it is necessary to describe how they can be concretely instantiated at run-time for the realization of a real open interaction system. One possible approach consists in proposing, coherently with the theory on agent environments, to instantiate AIs by introducing in the model of open distributed system the notion of *institutional space* of interaction [67, 36]. Institutional spaces are crucial because they allow to represent the boundaries of the effects of institutional and physical events or actions performed by agents, secondly they are the component in charge of enforcing the norms in response to the happened events/actions. Institutional spaces can be created and destroyed run-time on the basis of the agents' interactions.

An interesting aspect of the research on AI and environment is due to the fact that the same AI or different AIs may be instantiated in different institutional spaces. Those spaces may exists in *parallel* (inside the same container, like for example different auctions inside a marketplace), and they may also contain *sub-spaces* (like for example the space of a marketplace that contains different spaces each one corresponding to a running auction). Therefore, it is relevant to study the *inter-dependencies* between AIs (at design time) and between different institutional spaces at run time.

One possible approach for managing those interdependencies is described in the Multi-Agent Normative EnvironmenTs (MANET) meta-model [67]. It consists in introducing the notion of *observability* of an event outside the boundaries of the institutional space where it happen. For example a norm defined in a marketplace may regulate the actions that an agent is allowed to perform in an auction (which is a sub-space of the marketplace). Another useful functionality is the *notification* of events among parallel institutional spaces. In fact, for example a norm inside one auction may regulate the action of an agent on the basis of the role that the same agent play in another auction.

Another interesting study on the specification and reasoning on multiple institutions is [16]. In this paper the formal specification of single institution [14] is extended to multiple institutions. This by extending the notion of institutional power to perform an action inside a single institution to the possibility for another

	Roles	Norms	Constitutive Rules
ANTE	Two type of roles: institutional roles and generic roles subject to norms	One normative environment with a set of regulations, which checks whether agents follow the norms, applies correction measures, and enables the run-time establishment of new normative relationships	Institutional facts are connected to brute facts (mainly agent illocutions) through appropriate constitutive rules
OCeAN + MANET	Roles are labels defined by a given AI, at design time they are associated to norms and powers	Specification of norms at design time associated to roles and dynamic creation of instances of norms at run-time associated to specific agents	One special constitutive rule for performing institutional actions by means of declarations
EIs + EIDE	Specification of role subsumption, and two forms of compatibility among roles	There is the possibility of explicitly expressing norms as production rules that are triggered whenever an illocution is uttered, thus allowing the specification of regimented and not-regimented conventions	The are not basic institutional facts, there is a domain language used in illocutionary formulas and whose terms correspond with physical facts and actions

Table 1: Comparison of Artificial Institution Models - Part I

institution to be empowered to change directly the state of another institution.

2.5 Comparison of Artificial Institutions Models

In Table 1 and 2, it is schematically summarized the support given, by three relevant models of Artificial Institutions, to the specification of the various components and concepts described in the previous sections; a more detailed comparison can be found in [30].

Those models of artificial institutions have been used for the realization of different prototype systems for solving different type of problems. For example the

	Organization of the interactions	Communication Language	Implementation
ANTE	Different *normative contexts* are established at run-time	Agents are free to inter-act with any other agents with any language, illocutionary actions may be performed by agents towards the normative environment as attempts to create institutional facts	Jade FIPA-compliant platform and Jess rule-based inference engine
OCeAN + MANET	The activities are realized into different *institutional spaces* or in *physical spaces* of interaction	Commitment-based semantics for assertives, commissives, and directives communicative acts	OCeAN: Event Calculus or Java + Semantic Web Technologies; MANET: PROLOG and GOLEM environment framework
EIs + EIDE	The activities are realized into *scenes*, which are connected by *transitions* creating a *performative structure*	All communications between an agent and the institution are mediated by a *governor*, utterance is admitted if and only if it complies with the institutional conventions	Z specification language and an ad-hoc peer-to-peer architecture

Table 2: Comparison of Artificial Institution Models - Part II

OCeAN + MANET model has been used for modelling an e-Energy Marketplace [67] and the norms that regulate the Dutch-auction [36]. The EIs model + EIDE framework has been used for modelling and develop a regulated open MAS able to manage water demand [8], a MAS decision support tool for water-right markets [39], an open MAS for realizing an hotel information system [57], and EIs for running the Spanish Fish Market [19].

3 Survey on Modeling Organizations in MAS

Besides institutions, organizations have also obtained increasing attention from the MAS community in the last years. An organization can be seen as a set of entities and their interactions, which are regulated by mechanisms of social order and created by more or less autonomous actors to achieve common goals. These agents and their goals are interlinked by some form of organizational structure and in most MAS research this structure is seen as a means to manage the complex dynamics in open MAS.

Looking at MAS literature especially the OPERA [21], TROPOS [9], GAIA [72] and MOISE+ and ORA4MAS methodologies [45] are often cited[7].

All of the just mentioned methodologies have in comment that they are based on organizational structures as their cornerstones. As such they recognize that when modeling the interaction of agents within a (open) MAS, it is not sufficient to simply focus on the architecture of the agents as well as their (communicative) abilities, and that one cannot assume autonomous agents to act according to the needs and expectations of the system design [41].

Organizations "represent rationally ordered instruments for the achievement of stated goals" [63]. As such they are being used to achieve specific objectives, which are defined by the specification of a number of sub-goals that are related to the overall goal of the organization. Looking at real world organizations, in business environments, an organization must furthermore consider the environment it is located in and exhibit characteristics such as a certain degree of predictability, stability over time as well as a focus on the organization's goals and strategies. Traditionally (as early as August Comte (1798-1854)), organizations therefore are considered to have two dimensions (that one needs to think of when wanting to model them): a factual dimension and a procedural one [65]. Whereas the factual dimension focuses on the observable behavior of the organization - and thus takes a more high level view on its goals as well as output - the procedural dimension has its focus on the question on how that behavior of the organization is achieved. In the procedural dimension therefore the view shifts to the division of labor to roles, the determination of authority and power as well as the establishment of communication links [21]. [3, Fig. 1] provides an overview of different MAS organizational architectures and functional and procedural / static dynamic features they exhibit.

3.1 Modeling Organizations: Between Top-Down and Bottom-Up

Wanting to model an organization at the very end of the spectrum, there are two opposite ways of designing it: (1) establish the organizational design off-line beforehand (design-time) or (2) let the organization be grown on-line from the bottom up by its participating agents (run-time). In the first case, the agents have no say in the global aims of the society. In the second case, which is more favoured by

---

[7]For a general survey on organization works in MAS see [44] or [50] for example.

researchers working on open MAS, the agents are the key and their goals as well as negotiations between them result in organizations being dynamically formed. A simple example of such an organization is twitter for example. Whereas the overall role of twitter is somewhat set (communication platform in the wider sense), the content of the communication is completely up to the users (within certain legal bounds). Thus, twitter hash-tags are not pre-define, but they are emerging based as a result of people using (and other people copying) them.

One highly discussed question within the MAS community is what is required to enable agents to do the latter. Main discussion point thereby is how rigid structures need to be that agents can use to use or establish organizations. Whereas in some works (e.g. [23]) advocate very rigid structures at instruction level that do not allow agents to deviate from expected behavior, other approaches attempt to aim for more flexible systems where agents can reason about deviating from expected behavior. The aim of this more flexible approach thereby is to enable agents (and indirectly thereby the organizations they are are situated in) to adapt to changes and extensions to the environment or to allow for 'foreign' agents to join [38].

3.2 Organizational Structure

The just outlined differences in organizational design that MAS designers face, have also been studied in the traditional organizational theory research area, where the two types of organizations are typically distinguished for human organizations: mechanical (sometimes also called mechanistic) and organic organizations [59].

In mechanical organizations - that closely relate to the design-time idea of MAS organizational design - tasks are precisely defined in advance, and they are broken down into separately specialized parts. Real world example of these kind of organizations are typically manufacturing companies, but also there are other groups that benefit from mechanistic organization; like universities.

In mechanical organizations, there is a strictly hierarchical structure both between the parts, but also between the knowledge and reasoning processes within the organization. In mechanical organizations communication tends to be mainly vertical, i.e. from the top (typical a centralized role) to the bottom of the hierarchy. Typical examples human organizations the have this kind of setup are bureaucracies and matrix structures [3].

Organic organizations in contrast are following concept of growing the organization from the bottom up. As such the members of/agents within such an organization can collaboratively (e.g. in groups) redefine and adjust the tasks, sub-goals and roles related to the organization, thereby possibly changing the whole organization and its goals. In organic organizations, less levels of authority and control are being present and communication is mainly horizontal rather then vertical. Knowledge and task control also tend to be distributed and reaction time to changes in such organizations is said to be shorter. Typical examples of organic organizations are team structures and virtual organizations [3]. Good examples of this type of structure would be Google and the coveted positions that lie within the Facebook Corporation.

From the above, for a designer wanting to decide on what the most suitable structure is for the system that he aims to model, he need to answer several questions first:

- Can the goals and tasks be divided into independent, formalized and standardized sub-tasks? An if so, how to approach this best?

- Which of the tasks and sub-tasks have dependencies that need considering?

- Can tasks be grouped together and what are good means to group tasks (function, geographical location, client, process, etc.)?

- At what level have decisions to be made and controls to be set up?

- What kind of environment is the organization located in (open, closed, static, dynamic)?

- What is the line of reporting in the organization? Who has authority and what is the chain of command?

- What rules and formal processes are being required in the organization?

- What level of predictability is the organization to have?

Goal of answering these questions is to enable the modeler of the organization to determine the main organizational features in order to develop the initial design of the organizational structure of the organization independently from the final use of agent concepts[3].

As a note to the reader, though we presented organizational structure as a single term above, in the real world and as a consequence also in organizational models, it can be multi-dimensional and consider several structured aspects at the same time that all need to be respresented. Some of the once already having been mentioned are "authority", "communication", "delegation", "responsibility", "control", "decision-making", "power", etc. [41].

3.3 Roles

The difference between the two concepts of textitInstitutions and *Organizations* can be exemplified when looking at the notion of 'roles'. From an institutional point of view roles are typically studied in terms of the set of norms associated with them, whereas from an organizational point of view the focus tends to be on the roles as a position in an organizational structure.

As organizations are typically established with some form of goal that is executed by agents enacting certain roles in mind, from an organizational perspective focus tends to be on roles that contribute to the achievement of the overall goal, rather then on the specific actors performing the particular roles. Thus, although it is the agent's capabilities that allow him to perform a role in a certain way, most of the time for an organization it is irrelevant who performs the role as long as it is performed. Think of a restaurant for example which will have several employees with the job title "cook". For a customer it is not relevant which (group of) cook(s) actually cooked his dish, as long as the dish tastes good and is delivered on time. This so called "role-oriented" approach is advocated by many works on organizations in MAS, including for example [3, 47, 25, 21].

Grossi [41] argues that from the structural point of view a role is just a position in a structure, that is to say, a set of links, whereas from the institutional perspective instead, they can be seen as a set of norms. Following the argumentation in [41] these two set partially overlap w.r.t the properties they express for transition systems. Whereas roles as set of norms specify how the role can be enacted, deacted, and what kind of status the agents acquire by enacting the role; roles as set of links specify the status acquired by agents playing certain roles (while disregarding how that role can be enacted or deacted), specify the the activities (e.g., delegation or information) that can be executed while enacting the role and, possibly, also their mental effects on the interacting agents.

Recapping, both institutions and organizations specify what an agent ought to, is permitted to, or has the right to do as well as have means to specify the status an agent playing a certain tole has acquired. What has to be noted however is that there are differences in this status specification. Whereas institutions connect abstract activities and state of affairs (i.e., transition and state types) to concrete ones, this is not the case for organizations. In contrast these have the activities that can be executed by agents playing a certain role as their main focus of attention. Thus, whereas institutions consider how a certain role can be reached, organizations are taking a look at what can be done while playing a role.

3.4 Comparison of Artificial Organization Models

The below table gives an overview how some organizational models in MAS mentioned here relate to the above. The below is not a conclusive overview, but rather a short glimpse into the different organizational approaches. A more detailed overview can for example be found in [4].

4 Conclusions and Forward Looking

Openness, decentralization, and heterogeneity of software components are fundamental characteristics of distributed systems operating on the Internet and in particular in the World Wide Web. At least since the 1990s, models and experimental implementations of open, decentralized and heterogeneous systems have been the main concern of the area of Computer Science research on Multi-Agent Systems (MASs). More recently, those studies went on with the proposal of numerous conceptual models of institutions and organizations. It is therefore not surprising this long tradition of studies may represent a fundamental source of ideas and methods for developing Web-oriented applications. In particular the sub-area of MAS research known as NorMAS (Normative Multi-Agent Systems) has been concerned with modeling, monitoring, and enforcing norms and policies in open distributed environments, producing solutions that have already been empirically tested with success, although mainly in the context of academic prototypes.

In this chapter we have provided an introduction to the basic concepts of modeling organizations and institutions in MAS and gave pointers to the work that has been done in the various NorMAS communities already. We started out by looking into institutions and discussed fundamental concepts such as regulative and constitutive norms, as well as ACLs and the challenges arising from institutions being situated in an environment that can impact on the institution. Afterwards our focus shifted to organizations. After briefly detailing the basic differences between organizations and institutions, the focus of the chapter afterwards turned to modeling organizational structure as well as modeling roles in organizations.

Looking forward, we believe that times are ripe for adapting institutions and organizations MAS models and techniques for solving real-world problems that arise on the Internet. Indeed, the development of advanced Web applications is already providing significant examples of actual applications on which the capabilities of MAS solutions can be put to be tested, evaluated, and improved.

For example, such solutions may be relevant for the regulation of access to generic datasets on the Internet or to datasets in the Web of Data, provided that they are implemented coherently with currently available Web technologies [26, 52]. Organi-

	Top-down/Bottom-Up	Structure	Dynamics
OperA	In opera the organizational model can be define whereas the social and the interaction model are consequences of the agent interactions. The OMNI approach [20], an extension of OperA introduces normative aspects and translates norms from an abstract level (in which organizational statutes and values are defined to a procedural level (where norms are implemented).	OperA describes the desired behaviour of the society and its general structure by means of an organizational model, where roles, interactions and social norms are described.	Organizational dynamics are detailed using a social model (in which agents are assigned roles using social contracts that describe the agreed behaviour inside the society) and an interaction model (that described the actual behaviour of a society during its interaction).
Tropos	An organizational model is used that details the organizations main actors, goals and dependencies at design time. This is done on a level of agent patterns (for particular roles) that are assigned to organizational topologies. The definition of social rules or global rules that apply to the whole organization are not considered.	Several organizational topologies and roles within these structures are considered. The structures include for example bureaucracy, matrix structures ad virtual organizations.	Social agent patterns are assigned to organizational topologies at design time. At run-time the effects of this assignment are analysed based on the pre-set rules.
GAIA (with Organizational Abstractions)	Main organizational goals of the system and its expected global behaviour are specified at design phase. Based on this organizations are established which can be divided into sub-organizations if needed (each of which can have its own structure). The environment, roles, interactions and social rules are also pre-defined.	GAIA considers that a specific topology for the system will force the use of several roles that depend on the selected topology pattern.	GAIA uses the concept of role-enactment, where the roles are depending on the chosen topology. Thus a change in the topology can alter the roles and their enactment, but not the other way around.

Table 3: Comparison of Artificial Organization Models

zational and institutional models may be crucial also for the realization of automatic machine-to-machine exchange of datasets when norms/policies and may be institutional concepts can be used for expressing the licenses [40] or ad hoc contracts/agreements that regulate the access and use of those data.

However, considerable research is still necessary before this approach can be adopted by industry-level products solving realistic problems. Moreover a deep comparison with approaches proposed in other fields of research is required.

References

[1] Huib Aldewereld and Virginia Dignum. Operetta: Organization-oriented development environment. In Mehdi Dastani, Amal El Fallah-Seghrouchni, Jomi Hübner, and João Leite, editors, *LADS*, volume 6822 of *Lecture Notes in Computer Science*, pages 1–18. Springer, 2010.

[2] J. Lluís Arcos, M. Esteva, P. Noriega, J. A. Rodríguez-Aguilar, and C. Sierra. Engineering open environments with electronic institutions. *Eng. Appl. of AI*, 18(2):191–204, 2005.

[3] Estefania Argente, Vicente Julian, and Vicente Botti. Multi-agent system development based on organizations. *Electronic Notes in Theoretical Computer Science*, 150(3):55–71, 2006.

[4] Estefania Argente, Vicente Julian, Soledad Valero, and V. J. Botti. Towards an organizational mas methodology. In *Recent Advances in Artificial Intelligence Research and Development*, Frontiers in Artificial Intelligence and Applications, pages 397–404. IOS Press, 2005.

[5] Alexander Artikis, Marek Sergot, and Jeremy Pitt. Specifying norm-governed computational societies. *ACM Trans. Comput. Logic*, 10(1):1:1–1:42, January 2009.

[6] John Langshaw Austin. *How to do things with words*. William James Lectures. Oxford University Press, 1962.

[7] Guido Boella and Leendert W. N. van der Torre. Regulative and constitutive norms in normative multiagent systems. In Didier Dubois, Christopher A. Welty, and Mary-Anne Williams, editors, *Principles of Knowledge Representation and Reasoning: Proceedings of the Ninth International Conference (KR2004), Whistler, Canada, June 2-5, 2004*, pages 255–266. AAAI Press, 2004.

[8] Vicente J. Botti, Antonio Garrido, Adriana Giret, and Pablo Noriega. Managing water demand as a regulated open MAS. In Matteo Baldoni, Cristina Baroglio, Jamal Bentahar, Guido Boella, Massimo Cossentino, Mehdi Dastani, Barbara Dunin-Keplicz, Giancarlo Fortino, Marie Pierre Gleizes, João Leite, Viviana Mascardi, Julian A. Padget, Juan Pavón, Axel Polleres, Amal El Fallah-Seghrouchni, Paolo Torroni, and Rineke Verbrugge, editors, *Proceedings of the Second Multi-Agent Logics, Languages, and Organisations Federated Workshops, Turin, Italy, September 7-10, 2009*, volume 494 of *CEUR Workshop Proceedings*. CEUR-WS.org, 2009.

[9] Paolo Bresciani, Anna Perini, Paolo Giorgini, Fausto Giunchiglia, and John Mylopoulos. Tropos: An agent-oriented software development methodology. *Autonomous Agents and Multi-Agent Systems*, 8(3):203–236, 2004.

[10] Maiquel Brito, Jomi Fred Hübner, and Rafael H. Bordini. Analysis of the use of events and states as brute facts in modelling of institutional facts. In *Revised Selected Papers of the COIN 2013 International Workshop on Coordination, Organizations, Institutions, and Norms in Agent Systems IX - Volume 8386*, pages 177–192, New York, NY, USA, 2014. Springer-Verlag New York, Inc.

[11] Amit K. Chopra, Alexander Artikis, Jamal Bentahar, Marco Colombetti, Frank Dignum, Nicoletta Fornara, Andrew J. I. Jones, Munindar P. Singh, and Pinar Yolum. Research directions in agent communication. *ACM Trans. Intell. Syst. Technol.*, 4(2):20:1–20:23, April 2013.

[12] Amit K. Chopra, Fabiano Dalpiaz, Fatma Başak Aydemir, Paolo Giorgini, John Mylopoulos, and Munindar P. Singh. Protos: Foundations for Engineering Innovative Sociotechnical Systems. In *Proceedings of the 22nd IEEE International Requirements Engineering Conference (RE'14)*. IEEE, 2014.

[13] Amit K. Chopra and Munindar P. Singh. *Agent Communication*, volume Multiagent Systems, chapter 3. The MIT Press, 2013.

[14] Owen Cliffe, Marina De Vos, and Julian Padget. *Answer Set Programming for Representing and Reasoning About Virtual Institutions*, pages 60–79. Springer Berlin Heidelberg, Berlin, Heidelberg, 2007.

[15] Owen Cliffe, Marina De Vos, and Julian Padget. Specifying and reasoning about multiple institutions. In Pablo Noriega, Javier Vazquez-Salceda, Guido Boella, Olivier Boissier, Virginia Dignum, Nicoletta Fornara, and Eric Matson, editors, *Coordination, Organization, Institutions and Norms in Agent Systems II - AAMAS 2006 and ECAI 2006 International Workshops, COIN 2006 Hakodate, Japan, May 9, 2006 Riva del Garda, Italy, August 28, 2006*, volume 4386 of *Lecture Notes in Computer Science*, pages 67–85. Springer Berlin / Heidelberg, 2007.

[16] Owen Cliffe, Marina De Vos, and Julian Padget. *Specifying and Reasoning About Multiple Institutions*, pages 67–85. Springer Berlin Heidelberg, Berlin, Heidelberg, 2007.

[17] M. Colombetti. A commitment-based approach to agent speech acts and conversations. In *Proceedings of the Workshop on Agent Languages and Conversational Policies*, pages 21–29, 2000.

[18] Domenico Corapi, Marina De Vos, Julian Padget, Alessandra Russo, and Ken Satoh. Normative design using inductive learning. *Theory and Practice of Logic Programming*, 11:783–799, 2011.

[19] Guifré Cuní, Marc Esteva, Pere García, Eloi Puertas, Carles Sierra, and Teresa Solchaga. Masfit: Multiagent system for fish trading. In Lorenza Saitta Ramon López de Mántaras, editor, *Proceedings of the 16th Eureopean Conference on Artificial Intelligence, ECAI'2004, including Prestigious Applicants of Intelligent Systems, PAIS 2004, Valencia, Spain, August 22-27, 2004*, pages 710–714. IOS Press, IOS Press, 2004.

[20] V. Dignum, J. Vazquez-Salceda, and F. Dignum. Omni: Introducing social strcuture,

norms and ontologies into agent organizations. In *ProMAS*, volume 3346 of *Lecture Notes on Artificial Intelligence*, 2005.

[21] Virginia Dignum. *A model for organizational interaction: based on agents, founded in logic.* PhD thesis, University Utrecht, 2004.

[22] A. A. O. Elagh. On the Formal Analysis of Normative Conflicts. *Information & Communications Technology Law*, 9(3):207–217, 2000.

[23] Marc Esteva. *Electronic Institutions: from specification to development.* PhD thesis, Insitut d'Investigació en Intel·ligència Artificial, 2003.

[24] Marc Esteva, Juan A. Rodríguez-Aguilar, Josep Lluís Arcos, Carles Sierra, Pablo Noriega, and Bruno Rosell. Electronic Institutions Development Environment. In *Proceedings of the 7th International Joint Conference on Autonomous Agents and Multiagent Systems (AAMAS '08)*, pages 1657–1658, Estoril, Portugal, 12/05/2008 2008. International Foundation for Autonomous Agents and Multiagent Systems, ACM Press.

[25] Jacques Ferber, Olivier Gutknecht, and Fabien Michel. From agents to organizations: An organizational view of multi-agent systems. In *Agent-Oriented Software Engineering IV*, volume 2935 of *Lecture Notes in Computer Science*, pages 214–230, 2003.

[26] T. Finin, A. Joshi, L. Kagal, J. Niu, R. Sandhu, W. Winsborough, and B. Thuraisingham. ROWLBAC: Representing role based access control in OWL. In *Proceedings of the SACMAT*, pages 73–82, New York, NY, USA, 2008. ACM.

[27] Tim Finin, Richard Fritzson, Don McKay, and Robin McEntire. Kqml as an agent communication language. In *Proceedings of the Third International Conference on Information and Knowledge Management*, CIKM '94, pages 456–463, New York, NY, USA, 1994. ACM.

[28] N. Fornara and M. Colombetti. Specifying Artificial Institutions in the Event Calculus. In V. Dignum, editor, *Handbook of Research on Multi-Agent Systems: Semantics and Dynamics of Organizational Models*, Information science reference, chapter XIV, pages 335–366. IGI Global, 2009.

[29] Nicoletta Fornara. *Semantic Agent Systems: Foundations and Applications*, chapter Specifying and Monitoring Obligations in Open Multiagent Systems Using Semantic Web Technology, pages 25–45. Springer Berlin Heidelberg, Berlin, Heidelberg, 2011.

[30] Nicoletta Fornara, Henrique Lopes Cardoso, Pablo Noriega, Eugénio Oliveira, Charalampos Tampitsikas, and Michael I. Schumacher. *Agreement Technologies*, chapter Modelling Agent Institutions, pages 277–307. Springer Netherlands, Dordrecht, 2013.

[31] Nicoletta Fornara and Macro Colombetti. A commitment-based approach to agent communication. *Applied Artificial Intelligence*, 18:853–866, 2004.

[32] Nicoletta Fornara and Marco Colombetti. Operational specification of a commitment-based agent communication language. In *Proceedings of the First International Joint Conference on Autonomous Agents and Multiagent Systems: Part 2*, AAMAS '02, pages 536–542, New York, NY, USA, 2002. ACM.

[33] Nicoletta Fornara and Marco Colombetti. Defining interaction protocols using a commitment-based agent communication language. In *Proceedings of the Second In-*

ternational Joint Conference on Autonomous Agents and Multiagent Systems, AAMAS '03, pages 520–527, New York, NY, USA, 2003. ACM.

[34] Nicoletta Fornara and Marco Colombetti. Specifying and Enforcing Norms in Artificial Institutions. In Matteo Baldoni, Tran Son, M. van Riemsdijk, and Michael Winikoff, editors, *Declarative Agent Languages and Technologies VI*, volume 5397 of *Lecture Notes in Computer Science*, pages 1–17. Springer Berlin / Heidelberg, 2009.

[35] Nicoletta Fornara and Marco Colombetti. Representation and monitoring of commitments and norms using OWL. *AI Commun.*, 23(4):341–356, 2010.

[36] Nicoletta Fornara and Charalampos Tampitsikas. Semantic technologies for open interaction systems. *Artificial Intelligence Review*, 39:63–79, 2013.

[37] Nicoletta Fornara, Francesco Viganò, and Marco Colombetti. Agent communication and artificial institutions. *Autonomous Agents and Multi-Agent Systems*, 14(2):121–142, 2007.

[38] Paolo Giorgini, Manuel Kolp, and John Mylopoulos. Multi-agent architectures as organizational structures. *Autonomous Agents and Multi-Agent Systems*, 13, 2002.

[39] Adriana Giret, Antonio Garrido, Juan A. Gimeno, Vicente J. Botti, and Pablo Noriega. A MAS decision support tool for water-right markets. In Liz Sonenberg, Peter Stone, Kagan Tumer, and Pinar Yolum, editors, *10th International Conference on Autonomous Agents and Multiagent Systems (AAMAS 2011), Taipei, Taiwan, May 2-6, 2011, Volume 1-3*, pages 1305–1306. IFAAMAS, 2011.

[40] Guido Governatori, Antonino Rotolo, Serena Villata, and Fabien Gandon. One license to compose them all - A deontic logic approach to data licensing on the web of data. In Harith Alani, Lalana Kagal, Achille Fokoue, Paul T. Groth, Chris Biemann, Josiane Xavier Parreira, Lora Aroyo, Natasha F. Noy, Chris Welty, and Krzysztof Janowicz, editors, *The Semantic Web - ISWC 2013 - 12th International Semantic Web Conference, Sydney, NSW, Australia, October 21-25, 2013, Proceedings, Part I*, volume 8218 of *Lecture Notes in Computer Science*, pages 151–166. Springer, 2013.

[41] Davide Grossi. *Designing invisible handcuffs. Formal investigations in institutions and organizations for multi-agent systems*. PhD thesis, Utrecht University, 2007.

[42] Davide Grossi, John-Jules Ch. Meyer, and Frank Dignum. Classificatory aspects of counts-as: An analysis in modal logic. *J. Log. Comput.*, 16(5):613–643, 2006.

[43] Pascal Hitzler, Markus Krötzsch, and Sebastian Rudolph. *Foundations of Semantic Web Technologies*. Chapman & Hall/CRC, 2009.

[44] Bryan Horling and Victor Lesser. Using odml to model and design organizations for multi-agent systems. In Olivier Boissier, Virginia Dignum, Eric Matson, and Jaime Sichman, editors, *Proceedings of the Workshop on From Organizations to Organization Oriented Programming (OOOP 05)*, 2005.

[45] Jomi F. Hübner, Olivier Boissier, Rosine Kitio, and Alessandro Ricci. Instrumenting multi-agent organisations with organisational artifacts and agents. *Autonomous Agents and Multi-Agent Systems*, 20(3):369–400, 2010.

[46] Andrew J. I. Jones and Marek J. Sergot. A formal characterisation of institutionalised

power. *Logic Journal of the IGPL*, 4(3):427–443, 1996.

[47] Ivan J. Jureta, Stéphane Faulkner, and Pierre-Yves Schobbens. Allocating goals to agent roles during mas requirements engineering. In *Proceedings of the 7th International Conference on Agent-oriented Software Engineering VII*, pages 19–34, 2007.

[48] Elias L. Khalil. Organizations versus institutions. *Journal of Institutional and Theoretical Economics (JITE)*, 151(3):445–466, September 1995.

[49] H. Lopes Cardoso. *Electronic Institutions with Normative Environments for Agent-based E-contracting*. PhD thesis, Universidade do Porto, 2010.

[50] Luciano, Jaime S. Sichman, and Olivier Boissier. Modeling organization in MAS: a comparison of models. *1st. Workshop on Software Engineering for Agent-Oriented Systems (SEAS'05)*, October 2005.

[51] Fabio Marfia, Nicoletta Fornara, and Truc-Vien T. Nguyen. *Multi-Agent Systems and Agreement Technologies: 13th European Conference, EUMAS 2015, and Third International Conference, AT 2015, Athens, Greece, December 17-18, 2015, Revised Selected Papers*, chapter Modeling and Enforcing Semantic Obligations for Access Control, pages 303–317. Springer International Publishing, Cham, 2016.

[52] Truc-Vien T. Nguyen, Nicoletta Fornara, and Fabio Marfia. Automatic policy enforcement on semantic social data. *Multiagent and Grid Systems*, 11(3):121–146, 2015.

[53] Pablo Noriega, Amit K. Chopra, Nicoletta Fornara, Henrique Lopes Cardoso, and Munindar P. Singh. Regulated MAS: social perspective. In Giulia Andrighetto, Guido Governatori, Pablo Noriega, and Leendert W. N. van der Torre, editors, *Normative Multi-Agent Systems*, volume 4 of *Dagstuhl Follow-Ups*, pages 93–133. Schloss Dagstuhl - Leibniz-Zentrum fuer Informatik, 2013.

[54] Douglass C. North. Institutions, organizations and market competition. Economic history, EconWPA, December 1996.

[55] Sascha Ossowski, editor. *Agreement Technologies*, volume 8 of *Law, Governance and Technology Series*. Springer Netherlands, Dordrecht, 2013.

[56] E. Ostrom. *Governing the commons-The evolution of institutions for collective actions*. Political economy of institutions and decisions, 1990.

[57] Armando Robles P, B. V. Pablo Noriega, Marco Julio Robles P., Héctor Hernández T., Victor Soto Ramírez, and Edgar Gutiérrez S. A hotel information system implementation using MAS technology. In Hideyuki Nakashima, Michael P. Wellman, Gerhard Weiss, and Peter Stone, editors, *5th International Joint Conference on Autonomous Agents and Multiagent Systems (AAMAS 2006), Hakodate, Japan, May 8-12, 2006*, pages 1542–1548. ACM, 2006.

[58] Jeremy Pitt, Julia Schaumeier, and Alexander Artikis. *Coordination, Conventions and the Self-organisation of Sustainable Institutions*, pages 202–217. Springer Berlin Heidelberg, Berlin, Heidelberg, 2011.

[59] Stephen P. Robbins and Timothy A. Judge. Organizational behavior, 2017.

[60] J. R. Searle. *The construction of social reality*. Free Press, New York, 1995.

[61] John R. Searle. *Speech Acts: An Essay in the Philosophy of Language*. Cambridge

University Press, Cambridge, London, 1969.

[62] J.R. Searle. A classification of illocutionary acts. *Language in Society*, 5(1):1–23, 1976.

[63] Philip Selznick. Foundations of the theory of organization. *American Sociological Review*, 13:25–35, 1948.

[64] Murat Sensoy, Timothy J. Norman, Wamberto W. Vasconcelos, and Katia Sycara. OWL-POLAR: A framework for semantic policy representation and reasoning. *Web Semantics: Science, Services and Agents on the World Wide Web*, 12-13:148–160, April 2012.

[65] Jaime S. Sichman and Rosaria Conte. On personal and role mental attitudes: a preliminary dependence-based analysis. In *Advances in Artificial Intelligence, 14th Brazilian Symposium on Artificial Intelligence, SBIA '98*, number 1515, 1998.

[66] Munindar P. Singh. Agent communication languages: Rethinking the principles. *Computer*, 31(12):40–47, December 1998.

[67] Charalampos Tampitsikas, Stefano Bromuri, Nicoletta Fornara, and Michael Ignaz Schumacher. Interdependent Artificial Institutions In Agent Environments. *Applied Artificial Intelligence*, 26(4):398–427, 2012.

[68] Andrzej Uszok, Jeffrey M. Bradshaw, James Lott, Maggie R. Breedy, Larry Bunch, Paul J. Feltovich, Matthew Johnson, and Hyuckchul Jung. New developments in ontology-based policy management: Increasing the practicality and comprehensiveness of kaos. In *9th IEEE International Workshop on Policies for Distributed Systems and Networks (POLICY 2008), 2-4 June 2008, Palisades, New York, USA*, pages 145–152. IEEE Computer Society, 2008.

[69] Danny Weyns, Andrea Omicini, and James Odell. Environment as a first class abstraction in multiagent systems. *Autonomous Agents and Multi-Agent Systems*, 14(1):5–30, 2007.

[70] M. J. Wooldridge and N. R. Jennings. Intelligent agents: Theory and practice. *The Knowledge Engineering Review*, 10(2):115–152, 1995.

[71] Pinar Yolum and Munindar P. Singh. Flexible protocol specification and execution: Applying event calculus planning using commitments. In *Proceedings of the First International Joint Conference on Autonomous Agents and Multiagent Systems: Part 2*, AAMAS '02, pages 527–534, 2002.

[72] Franco Zambonelli, Nicholas R. Jennings, and Michael Wooldridge. Developing multiagent systems: The GAIA methodology. *ACM Trans. Softw. Eng. Methodol.*, 12(3):317–370, 2003.

Received 8 March 2017

Modeling Norms Embedded in Society: Ethics and Sensitive Design

Rob Christiaanse

Delft University of Technology, The Netherlands, and EFCO BV, Amsterdam, The Netherlands
r.christiaanse@efco-solutions.nl

1 Introduction

1.1 Acting or not acting

When we start to think about moral norms, morality if you like, we encounter very intriguing problems about situations in which people find themselves facing a choice to make which are of excessive complexity. Take for example "The Trolley problem": "The trolley rounds a bend, and there come into view ahead five track workmen, who have been repairing the track. The track goes through a bit of a valley at that point, and the sides are steep, so you must stop the trolley if you are to avoid running the five men down. You step on the brakes, but alas they don't work. Now you suddenly see a spur of track leading off to the right. You can turn the trolley onto it, and thus save the five men on the straight track ahead. Unfortunately there is one track workman on that spur of track. He can no more get off the track in time than the five can, so you will kill him if you turn the trolley onto him. Is it morally permissible for you to turn the trolley" [70]. Or the problem addressed by Macintyre in [45] where he starts off with the case of J (who might be anybody, *jemand*) analyzing whether J's defense to the allegation of moral failure holds because J failed his responsibility.

1.2 Using (il)legitimate evidence

Let's look into some judicial cases. Regulators use information obtained from different sources to assess whether businesses and citizens comply with applicable laws and regulations. Recently the supreme court of the Netherlands ruled in case number [14]. The key issue addressed in this case was whether the Tax authority was allowed to use information obtained from Automatic Number Plate Recognition

(ANPR) systems checking tax returns from employees driving company leased cars. In defence the defendant plead that article 8 of the ECHR protects ones private live and therefore the tax authority was not allowed to use the aforementioned information to check the tax return as done. The Supreme Court ruled that in this case it was wrongly assumed that the general job description of the tax authority or any (other) provision provides in a sufficient basis to use information from Automatic Number Plate Recognition (ANPR) systems. In 2015 a court procedure was ruled in a similar case [13]. A vehicle driver was fined for speeding on the A2. The A2 is a motor highway in the Netherlands equipped with a section control system using ANPR for detecting all vehicles driving on the A2. The section control system measures the time and records a timestamp image of the license plate from the moment the car enters the section until the car leaves the section. During the route (i.e. the section) the process of measuring en recording is repeated at fixed intervals. After the car has left the section the actual speed is calculated and in the case the average speed exceeds some threshold the vehicle driver is fined automatically. In this case the car driver drove 8km/hr. to fast and got fined for EUR45. The man went to court and stated that the section control system infringed his private life so article 8 of the ECHR was violated. In this case the judge ruled that the police law provided in a sufficient basis to use the information obtained from the section control system. As for the claim that the fined man's private life was infringed the judge did not accept this line of reasoning because all vehicle drivers are well informed of the system by means of traffic signs informing car drivers that speed is measured by means of a section control system. Additionally the records of the time stamp images of the license plates are to be disposed of after 72hours counting from the moment the recording of timestamp images starts. Considering the aforementioned circumstances the judge ruled that the privacy concerns were only violated on a limited level. The fine had to be paid by the car driver.

Both cases share that information obtained from ANPR systems exist and that regulatory bodies use this information as evidence for regulatory oversight purposes. Both share the issue whether usage by regulatory bodies of this evidence infringe fundamental rights that protects ones private live. At first sight the reasoning of the judge in the first case is quite straightforward namely in the first case the "job description" did not provide in a sufficient basis for using the specific information for regulatory oversight purposes so the issue whether the usage of the specific information infringed fundamental rights that protects ones private live need not to be addressed. The nature of the ruling is what we will coin as formal procedural. In the second case the judge ruled that the usage of the specific information did fit the job description and that usage of the specific information was granted and therefore legitimate so the issue whether the usage of the specific information infringed the fundamental rights had to be addressed. Due to the fact that the vehicle driver was well informed at the time he or she drives onto the motor highway A2; he or she should be aware that the car he or she is driving will be under surveillance of the section control system. Due to the fact that the records are only kept for a limited time period the judge reasoned that the fundamental rights preserving ones private live was violated on a limited level, therefore the usage of the specific information for regulatory oversight purposes was legitimate. Hence the judge weighted the consequences for the parties involved. The nature of the ruling is both formal procedural and substantial.

1.3 Ethics and moral

In the case we have to model norms embedded in society with an ethical and sensitive design in mind than the inevitable and pressing question is: "what is an ethical and sensitive design in the first place?" Ethics is a branch of moral philosophy, addressing questions like: "what is a wrong thing to do and what is a right thing to do (in situation x for example) when y happens". In the cases we introduced in the former section, we simply have to ask ourself in simple present future tense "is it a right thing to do?" or "is it a wrong thing to do" or in past perfect tense "was it a right thing to do?" or "was it a wrong thing to do". Wrong is not the opposite from right and right is not the opposite from wrong. In moral theory questions about value play a major role. In a very narrow sense value theory refers to axiology addressing questions whether objects of value are psychological states or objective states of the world. Put in a more broader context the value theory concept addresses questions about the nature of value and its relation to other (moral) categories like naturalistic goods opposed to human made entities i.e. artifacts. With this distinction in mind we get a mechanism enabling us to reason about values and tradeoffs, often coined as an evaluation mechanism. In the above illustrated cases we recognise that values are weighted but that the underlying mechanism is often opaque by nature and therefore hard to decipher. How these trade-offs are made is less understood for example when pluriformity in moral values may exist.

1.4 Judgement

Modelling norms implies that we have to deliberate about the nature of norms to model in the first place. The concept of value seems a intertwining concept among alternatives to choose from. In some cases a tradeoff has to be made by some human or technological component if you like. Making trade-offs brings in the question of agency. Agency is the capacity to make choices to act [45] [50]. Moral agency is the

ability to make moral judgements based on some notion of right or wrong . Hence
there exists some evaluation function to judge. Indeed the evaluation function is a
necessary condition, whether the evaluation function is sufficient depends on what
is believed to be true and justified. Recognise the epistemic nature of the logic
buttressing the evaluation function. Moral responsibility on the other side is about
human action and its consequences. The concept of responsibility can be viewed
from two distinct though interrelated concepts namely (1) the merit-based view and
the (2) consequentialist view [68, 69]. Broadly speaking the distinction draws the
line between responsibility as accountability versus responsibility as attributability.
Attributability is a necessary but not a sufficient condition for being accountable. In
the case humans have to deal with with moral issues in certain situations, the concept
of compartmentalization seems important. "Compartmentalization goes beyond that
differentiation of roles and institutional structures that characterizes every social
order and it does so by the extent to which each distinct sphere of social activity
comes to have its own role structure governed by its own specific norms in relative
independence of other such spheres (social space). Within each sphere those norms
dictate which kinds of consideration are to be treated as relevant to decision-making
and which are to be excluded" [45].

1.5 Eliciting requirements in making social and moral values to design

In today's society individuals and institutions act with and in sociotechnical systems. Humans and technological components interact with each other and affect
each other in contingent ways. Designers of aforementioned sociotechnical systems
face a tremendous task in how to address moral norms and how to elicit requirements
to impose onto the design, built and implementation of a sociotechnical system. The
type of sociotechnical systems we have in mind form the class of normative multi
agent systems as defined in [2] Making social and moral values central to a design
stems from the 1970s at Stanford often referred to as Value Sensitive Design (VSD)
first formulated by Friedman [20, 21, 22, 23, 36]. Many similar approaches followed
coined as "values in design", "values and design" and "design for values " [36]. VSD is
a reaction to the idea and practice that a design of an artifact whatever that may be
is a foremost technical and value-neutral task primarily focused on the requirements
of users of the artifact. VSD is a theoretically grounded approach to the design of
technology that accounts for human values in a principled and comprehensive man-
ner throughout the design process[21]. Artifacts as such are the result of thousands
of design decisions. The fact is that these decisions may affect one's health, safety,
identity or society at large compare coined as "Diesel Dupe" where the EPA (envi-

ronmental Protection Agency) detected a "defeat device" or software intentionally designed to cheat with the emission tests in the US. It is the decision to design a "cheat devise" that makes the issue morally questionable. What we would like to point out is that design processes are value sensitive in nature and that it is the choice of the designers whether ethical aspects should be taken into account in the design processes. Therefore we need an explicit interpretation of what is constituted as the tacit understanding, just displayed i.e. showed in practice[34];"testing a design hypothesis is inextricably bound up with the ethical normative framework of society and with its epistemological principles"[18]. Modeling norms embedded in society in an ethics and sensitive design perspective is not about modeling ethics or moral reasoning but reflects the decisional processes buttressing a design process taking ethical or moral reasoning into account. Indeed a value design perspective is concerned with the mechanisms making a design process ethical and morally sound or unsound. Much of the debates concern the development of information systems technology. Floridi formulated eighteen open problems in the Philosophy of Information [16] covering fundamental areas like the information definition, information semantics, intelligence/cognition, informational universe/nature and values/ethics. The key question, question P18 is: "does computer ethics have a philosophical foundation?" These types of questions are distinct from the value sensitive design perspectives. The question of philosophical foundation relates to the uniqueness debate [17].

1.6   outline

In chapter 2 we start with the introduction of a code of conduct implemented by Nike [53, 54]to frame the design task in a principled approach. In this chapter we address the notion of moral values using categories of type of ethics. First we distinguish descriptive ethics from normative ethics which forms can either be rule-based or virtue-based. Next we address the foundational aspects of any type of ethics, value pluralism and decision procedural aspects to come up with a procedure to classify and analyze characteristics of an ethical system. Ethics is always personal to a human, and in the case a situation is the consequence of human decision making, persons may be under a duty to apply value judgments to the consequences of their decisions, and held responsible for those decisions. Reflexivity shape norms, tastes and wants of an agent and determine the effectiveness of any system. After elaborating on the notions of decision rights, responsibility and accountability we end up in this chapter with rephrasing the original design question into 7 key questions formulated in a principled way using the procedure to classify and analyze an ethical system applied to the code of conduct op Nike. In chapter 3 we address the notion of a model as the start and the result of the design process. Chapter 4 entails some concepts and definitions buttressing normative multi agent systems. Especially we address the problem of mechanism design as formulated by [38]. Using a verification scenario which separates the process of finding an equilibrium from recognizing an equilibrium it is possible to design incentive compatible mechanisms which occurs when the incentives that motivate the actions of individual participants are consistent with following the rules established by the group. This notion is paramount in establishing whether the mechanism designed is indeed effective. We end up in chapter 4 with some observations and characterizations of a design in general. In chapter 5 we explore the notion of relationships and values and the commonality of exchange mechanisms when studied from a sociological, anthropological or economical point of view. We analyze an interaction model between two agents addressing the question: How much is either agent willing to "give up" (i.e. to sacrifice) to "get" (i.e. to gain) the wool respectively the cloth? Key problem addressed is how models representing exchange mechanisms can be extended with notions of measurement and valuation. Chapter 6 we elaborate on principles, architectures and state transitions systems. Models are analogous to Janus structures representations with an engineering side facing the real world and an abstract side facing theories[67]. In design practice it is simpler to formulate theories in first order logics and use explicitly meta reasoning about axioms and postulates; known as the AGM axioms for theory revision [1]. Finally in chapter 7 we elaborate on ethical sensitive design. First we reflect on the decision right allocation procedure and the verification mechanism. Secondly we introduce the notion of creating a vision from first principles.

2 Moral value(s)

Moral values play a crucial role in our society at large. In the case we were asked to list examples than the list of examples would be endless. In everyday life we all experience issues somehow, direct or indirect, related to norms, morality, and ethical behavior. Can we clarify what is meant by moral values? We start with a short description of a real life example. Nike is required by the The California Transparency in Supply Chains Act of 2010 (SB 657) ("Act") effective from January 1 2012 in the State of California to disclose efforts to eradicate slavery and human trafficking from direct supply chains [53, 54]. Nike raises expectations of their factory partners through standards written down in a Code of Conduct containing a statement of values, intentions and expectations meant to guide decisions in factories. The Code of Conduct of Nike is freely available for the public and expresses on merit grounds what is expected of factory partners [54]. We site some parts of the text.

- understanding that our work with contract factories is always evolving, this Code of Conduct clarifies and elevates the expectations we have of our factory suppliers and lays out the minimum standards we expect each factory to meet

- It is our intention to use these standards as an integral component to how we approach NIKE, Inc. sourcing strategies, how we evaluate factory performance, and how we determine with which factories Nike will continue to engage and grow our business

- We believe that partnerships based on transparency, collaboration and mutual respect are integral to making this happen

- Our Code of Conduct binds our contract factories to the following specific minimum standards that we believe are essential to meeting these goals

Next there are eleven "principles" listed such as:

- EMPLOYEES are AGE 16 or OLDER. Contractor's employees are at least age 16 or over the age for completion of compulsory education or country legal working age, whichever is higher. Employees under 18 are not employed in hazardous conditions

- HARASSMENT and ABUSE are NOT TOLERATED. Contractor's employees are treated with respect and dignity. Employees are not subject to physical, sexual, psychological or verbal harassment or abuse

- The CODE is FULLY IMPLEMENTED. As a condition of doing business with Nike, the contractor shall implement and integrate this Code and accompanying Code Leadership Standards and applicable laws into its business and submit to verification and monitoring. The contractor shall post this Code, in the language(s) of its employees, in all major workplaces, train employees on their rights and obligations as defined by this Code and applicable country law; and ensure the compliance of any sub-contractors producing Nike branded or affiliate products

2.1 Design question

Suppose hypothetically that we were asked to design a normative multi agent system based on the Code of Conduct of Nike. We can come up with a design question as follows: "why is a code of conduct needed?" and "does the contents of the Code of Conduct meets its objectives of the public, here the state of California?" Subsequently the next question emerges: "if we implement the current code of conduct will it suffice to ensure Nike and other stakeholders of the company that the suppliers actually realize the objectives as stated in the code of conduct?" A normative multi-agent system can be defined as a system by means of mechanisms to represent, communicate, distribute, detect, create, modify, and enforce norms, and mechanisms to deliberate about norms and detect norm violation and fulfillment [2]. We adopt the mechanism design definition because is formulates precisely what a normative multi-agent system does. Alas our definition does not help us right away how to elicit the norms themselves. We have to look at the design question more in-depth and take a closer look at the content of the code of conduct.

2.2   Descriptive ethics

As we learned Nike is required by the "The California Transparency in Supply Chains Act of 2010 (SB 657) ("Act")" effective from January 1 2012 in the State of California to disclose efforts to eradicate slavery and human trafficking from direct supply chains [53]. Apparently this is why we need a code of conduct. Code of conducts come in various forms and are most of the times descriptive in nature: in our example it discloses efforts to do x to achieve y. By means of the code of conduct stakeholders can uncover management's attitude, convictions and conceptions towards values that matter. Indeed it reflects what actions society rewards or punishes; in our case in the first place the law. Descriptive ethics involves empirical investigation often studied in the fields of biology, psychology, sociology, economics and management sciences. Theories and empirical findings find their way in philosophical arguments. As a consequence descriptive ethics is relativist, situational, situated or both. Merely descriptive ethics relate to the discourse of social sciences i.e. cultures and cultural norms: conceived as standards of proper or acceptable behavior [49]. Culture can be defined as "Culture is a patterned way of thinking, feeling and reacting, acquired and transmitted mainly by symbols, constituting the distinctive achievements by human groups, including their embodiments in artifacts. the essential core consists of ideas and especially their attached values."[42].

2.3   Normative ethics

The state of California aims to eradicate slavery and human trafficking from direct supply chains. In this context to eradicate means to destroy, exterminate practices like slavery and human trafficking in supply chains. In the case we interpret de Law normatively than our analysis will not start with the code in conduct in mind, but we would start to ponder about how one ought to act. Normative ethics studies ethical action, more precisely what makes actions wrong or right. Hence the fact that there is a Law stating that slavery and human trafficking has to be exterminated means that there is a practice of slavery and human trafficking; and that the institutions of the state of California has judged that slavery and human trafficking is no longer accepted by society and therefore such practices are no longer accepted. Using the term eradicate does not make a Law normative in nature in the deontological sense. Where the meaning of moral language is concerned lies in the realms of meta-ethics. A meta-ethical question would be "is is possible to eradicate slavery and human trafficking in supply chains in general". Positing this question introduces important notion whether an ethical claims can be judged at all. Deontology is the study of that which is an "obligation or duty", and consequent moral judgment on the actor on whether he or she has complied. Deontology is an approach to ethics that determines whether acts, or the rules and the duties of an agent performing the act is good or right. So goodness or rightness is judged by the act itself and not by the consequences. In deontology it is possible that an act considered as right or good that the act itself produces bad consequences even (!) in the case an agent who performed the act lacks virtues and had a bad intension in performing the act. The same is true in the case an act does not have any consequences at all in terms of the resulting effect pursued by performing the act. Hence it is possible that an agent did perform the act purposely wrong than he or she still did the right thing in deontology approaches to ethics. In the case we want to rule out such behavior than we have to spell out all acts, rules and duties to comply with. Up to this point we stress to point out that there is a distinction between rule-based ethics versus virtue-based ethics. Like in deontological approaches to morality the rule-based ethics focuses on acts and maintains that these rules are moral or not, to the extent of conformity, failures to conform to certain rules or principles. Virtue ethics on the other side holds the position that morality rests upon moral qualities. So it seems correct to impose that rule-based ethics is governed by concepts like acts, moral rules and moral principles and that virtue ethics is governed by moral dispositions, emotions, states of character and the flourishing of human beings. In virtue ethics morality is directly linked i.e. intimately linked to the person who acts, to his or her character and situation. It can be the case that the same act is morally wrong or right subsidiary on who acts and the conditions under which the act is done [60, 61]. We recognize the concept of compartmentalization stating that differentiation of roles and institutional structures that characterizes every (social) order and it does so by the extent to which each distinct sphere of social activity comes to have its own role structure governed by its own specific norms in relative independence of other such spheres (social space). Within each sphere those norms dictate which kinds of consideration are to be treated as relevant to decision-making and which are to be excluded. [45].

2.4 Many values

We could extent our analysis for all categories of ethics like anarchist ethics, pragmatic ethics, role ethics, information ethics, machine ethics, utilitarian ethics, virtue ethics, hedonism, consequentialism, stoicism, evolutionary ethics, applied ethics like business ethics et cetera. An ontological approach seems to fail in getting answers to our design questions in the case we were hypothetically asked to design a normative multi agent system based on the Code of Conduct of Nike. Simply because we have to evaluate whether one of the type(s) of ethics fits our purpose so we can determine design principles which guides our design in setting, implementing and realizing the design objective(s) (i.e. goal(s)). A design principle is a normative principle on the design of the artifact [31]. In general ethics or moral philosophy is a branch of philosophy that deals with values relating to human conduct, with respect to rightness (goodness) and badness (wrongness) of motives and end(s) of such actions. It needs no elaboration to state that there are many different moral values coined as moral value pluralism. Hence value pluralism is not about different value systems. Our key concern is how to evaluate all these different values in a coherent way, so we can make informed decisions to characterize the ethics adhered by stakeholders. Moral values can be characterized by being monist or pluralist[46] denoted as **M** respectively **P**. Monists claim that there is only one ultimate value. Pluralist defend the position that there are several distinct values. Each value can be classified along three dimensions namely (1) foundational, (2) normative and (3) decision procedural. Foundational entails that there is no one value that subsumes other values denoted as **F**. Normative posits there is a bearer of value denoted as **N**. The third characteristic decision procedural refers to a certain form of consequentialism (i.e. a representation) has its criterion of goodness or the right action denoted as **S**. The possible relationships are represented in figure 1 value pluralism.

Now we can come up with a procedure to classify a moral value and analyze the characteristics. Reading from left to right we are able to generate a table with all possibilities.

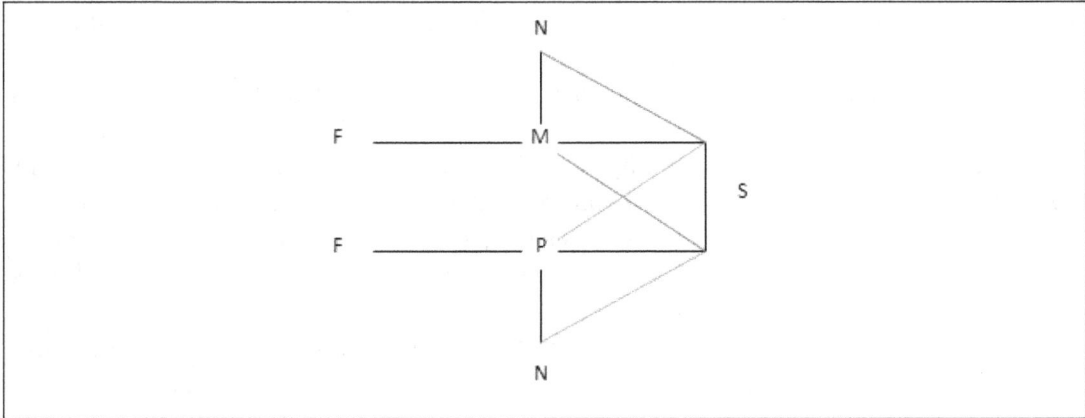

Figure 1: Pluralism

	Found	Mon/Plu	Norm	Scale
	F	M	N	Representation
	F	M	¬N	Representation
	F	P	N	Representation
	F	P	¬N	Representation
Moral Value				
	¬F	M	N	Representation
	¬F	M	¬N	Representation
	¬F	P	N	Representation
	¬F	P	¬N	Representation

Observe that it does not matter how the table is scrambled. Consider the features of **F**, **M**, **P** and **scale** as categories in the set theoretical sense. Observe that the set of scales can be empty. Informally we are now able to define a moral value (MV) as a set MV $=((F),(N),((P),(M)),(S,\emptyset))$.

2.5 Human agency

Human agency is the capacity to make choices and entails the claim that humans do in fact make decisions and enact them on the world [50]. This particular capacity is always personal to that human, though considerations of the outcomes enacted from private acts of human agency for us and others can then be thought of as an instantiation of moral value of a given situation wherein agents will, or have acted. In this type of situation we speak of moral agency. If a situation is the consequence of human decision making, persons may be under a duty to apply value judgments to the consequences of their decisions, and held to be responsible for those decisions. To understand moral value is to understand how decision rights are dispersed among agents and who decides. Note we are not addressing how agents decide or how they come to decisions. Discussions on the notion of free will and theorizing on the nature of rationality in making choices et cetera are very important to understand the effectiveness of a (moral) value system but it does not affect the question who decides justifying the choices buttressing a (moral) value system. Since the capacity of decision making is always personal to begin with, we have to address an important concept known as reflexivity. Reflexivity has a profound place in social theory and refers to an act of self reference recognizing forces or pressure within the environment and his or her place in the social structure. Agents with a low level of reflexivity are said that the environment shapes the individual norms, tastes, wants et cetera. Agents with a high level of reflexivity shape their own norms and tastes. Reflexivity addresses autonomy and thus autonomous action of an agent. Reflexivity is both a subjective process of self-consciousness inquiry and the study of behavior where relationships are concerned. Reflexivity seen as a subjective process of self-conscious inquiry is phenomenological in nature. Phenomenology studies the structures of consciousness as experienced from the first persons perspective. Here we enter the realms of intentionality, being the central structure of an experience directed towards an object by virtue of its content or meaning which represents the object [5]. We will address the meaning of intentionality later on in more detail. Agency and reflexivity play an important role in designing effective and efficient (corporate) organizational structures, information and communication systems [57]. It needs no elaboration that with decision rights and the choice to exercise these decision rights (i.e. often coined as (decision) power) there comes a responsibility issue. Decision rights, responsibility and accountability are intertwined concepts hard to decipher. One can feel responsible and act accordingly what the agent sees fit at some moment, although the agent did not have any formal decision rights attributed to him. Did the agent do the proper thing? If the situation turned out to be worsened because of the agent's interference, how would we judge? On the other hand an agent can decide for whatever reasons not to act, although the agent did have formal decision rights attributed to him. Did the agent do the proper thing? If it turns out that it was a bad decision not to act than we might judge de agent to be negligent. How to reason when somebody delegates his or her decision task to another agent and the same situations occur under the condition of delegation? The concept of responsibility can be viewed from two distinct though interrelated concepts namely (1) the merit-based view and the (2) consequentialist view. Following Strawson the distinction draws the line between responsibility as accountability versus responsibility as attributability [69]. The two categories do not always fit the situation. Contemporary views have introduced what is coined as "The answerability model". In this view attributability and accountability selfdisclosure is the target of appraisal and is judgemental sensitive. Indeed reflexive notions can play a role here, like socialisation and adaptive behavior et cetera. Here we have the proper considerations and motivation for the need of normative multi agent systems as we defined it above. Moral behavior of agents needs to be monitored and outcomes are judgmental sensitive if we accept a human centered perspective on systems.

2.6   Design question revisited

If there is a moral value than there must be a belief characterized as a propositional attitude, informally defined as the mental states of an agent or a group of agents having some kind of attitude, or opinion about a proposition or about a potential state of affairs in which a proposition is true. In our case it is the belief of the state of California that slavery and human trafficking should be rooted out in the supply chain of companies resident in the state of California. Needless to say that a belief characterized as a propositional attitude is similar to the goal setting processes of enterprises, organizations, forms, soccer clubs et cetera, sharing their vision by means of belief systems [66]. Furthermore there is a decision right allocation procedure buttressing responsibilities and accountability and warrants that the appropriate rules, standards, regulations, rewards and punishment are established. We extent our definition **MV** with beliefs denoted as B and decision right allocation procedure denoted as **DRAP**; we informally define MV = ((B),(F),(N),((P),(M)),(S,$\emptyset$),(DRAP)).

Now we can return to our design question. We asked ourselves "Why is a code of conduct needed?" and "does the contents of the Code of Conduct meets its objectives of the public, here the state of California?" Subsequently the next question emerged: "if we implement the current code of conduct will it suffice to ensure Nike and other stakeholders of the company that the suppliers actually realize the objectives as stated in the code of conduct?" By rephrasing the original questions we get 7 key questions to address:

	Questions	Sets
1.	What is the believe of the state of California with reference to slavery and human trafficking?	$\mapsto$B
2.	Are the values expressed by extricating slavery and human trafficking from direct supply chains subsumed in other values?	$\mapsto$F
3.	Are there several distinct values expressing extricating slavery and human trafficking?	$\mapsto$P,M
4.	Who are the value bearers in the supply chain?	$\mapsto$N
5.	How are the decision rights dispersed in the supply chain, who is responsible and accountable ?	$\mapsto$DRAP,N
6.	What rules, standards, regulations, rewards and punishment are established preserving moral values in the direct supply chains?	$\mapsto$N
7.	If applicable is there a representation expressing the moral value in communication processes?	$\mapsto$S

Observe that the object in the case of Nike is the direct supply chain. So from a design perspective the design objective is to come up with **a design** for direct supply chains that warrants that slavery and human trafficking is rooted out from the direct supply chains of Nike that is of the production of sportswear. The seven questions guide the designer to elicit the informational requirements. When we humans design things for the purpose of improving thought or action we actually create an artifact that has a physical presence we can actually manufacture or construct or has a mental presence we can actually learn. Both artifacts are equally artificial since they both would not exist without human invention. Clearly cars, papers, computers, doors, sportswear et cetera are physical artifacts, where reading, listening, logic, language are mental artifacts. Hence mental artifacts produce rules and structures in information structures. It needs no elaboration that mental and physical artifacts are related to each other. Consider in our case the distinction between rule-based ethics versus virtue-based ethics. Clearly rule-based ethics and virtue-based ethics are mental In the case we accept that physical and mental artifacts are equally artificial than we must accept in the real that information structures represented in the mental artifact are equivalent to the physical artifact represented in terms of physical properties[55]. This is why representations express any system; the notion of model plays a central role in any design.

3   The notion of a model

A model is considered to be a representation of some object, behavior, or a system that one wants to understand [55]. In everyday life we are on a day to day basis involved in making decisions about what should a model should look like to become meaningful and therefore useful. For example think of working procedures at your work, deadlines to meet with clients, appointments to make for personal reasons at the doctor, working conditions to respect when we engage in a trade in foreign or domestic countries, appraisal to give in the case good work is delivered by an employee, or punish someone who didn't do a proper job et cetera. In all these cases we have some sort "workflow model", "process model" or "communication model" in mind shared with colleagues, clients, vendors, et cetera in some format, which for example can serve as a plan for organizing ones work, getting things done, or is somehow useful for goalsetting purposes or all three together. In the end you want to make sure that people you work and communicate with understand your goals and your wants. A model is always the result and the start of a design process. "A design process is an abductive sensemaking process, a step of adopting a hypothesis as being suggested by the facts ... a form of inference, albeit inference of "best guesses" leaps [..]. *A logic of what might be.* It is not entirely accurate, âĂę it is the argument to the best explanation, the hypothesis that makes the most sense given observed phenomena or data and based on prior knowledge" [43]. In a scientific context testing a hypothesis means confronting statements about an assumed relationship between phenomena with empirical facts. In a design context the terms testing and hypothesis tends to shift in meaning, a design assumption is not a matter of being true or false but given a particular context it is a matter of being the best solution based on vision and believes [10]. As a consequence very different logics of discovery may be at work in design practices, and the way they are mixed may vary form case to case, from situation to situation, from context to context and so on. Whatever mix or configuration of elements, we will always need (good) theories to account for what happened. Theories are constitutive for every design just because we need to understand why some things do work and other things do not work or will never work. We must explain in advance why. Put in other words, we need a explicit interpretation of what is constituted as the tacit understanding, just displayed i.e. showed in practice [34]. Some authors state that "testing a design hypothesis is inextricably bound up with the ethical normative framework of society and with its epistemological principles" [18]. A model represents a justified true believe. A specific issue has to be addressed know as the Gettier problems in epistemology [25]. There are two generic features that characterize the original Gettier cases, (1) fallibility and (2) luck. Fallibility implies that there are strong indications that the justification favors that the belief is true, without proving conclusively that it is. Luck refers to how the belief manages to combine being true with being justified given the fact that the well but fallibly justified belief in question is true. This notion is very important in the case we have to gather evidence and make inferences whether the behavior of a system, group of systems, human, groups of humans, object or a group of objects can be judged to behave ethical or being ethical.

4 Normative multi-agent systems: some concepts and definitions

4.1 Agents

An agent is defined as "a computer system that is situated in some environment, and that is capable of autonomous action in this environment in order to meet its design objectives"[72]. An intelligent agent is such a computer system that is capable of flexible autonomous action where flexibility implies (1) reactivity, (2) pro-activeness and (3) social ability. Reactivity means that the computer system is aware of its environment, and is able to respond in a timely fashion to changes. Pro-activeness means that the computer system is able to take initiative. The third implication concerns the ability to interact with other computer systems and humans. A multi-agent system is a system composed of multiple, interacting computer systems. What makes a multi-agent system intelligible? Imagine that only one computer system has the capability of flexible autonomous action and the other computer systems do not have this capability? What type of agent system do we have? Hence identifying that some system is hybrid is not enough. We need to know precisely. In [73] the definition of an agent is slightly altered. Instead of design objectives, agents should meet delegated objectives. This seems trivial from a machine centered point of view but from a human centered point of view this is a big shift in perspective and has major implications. The notion of delegation is directly related to the notion of agency, accountability and responsibility. If it is possible to design, build and implement environmental aware computer systems, and these computer systems become or are empowered than studying multi-agent systems from a normative stance raises some deep fundamental theoretical issues with large practical implications. One should bear in mind for example that empowerment in general beholds a management practice of sharing information, rewarding personnel, and share decision power with employees so that they can take initiative and make decisions to solve problems of some kind to improve services and performance [65]. Empowerment is based on the idea that giving employees capabilities, resources, authority, opportunity, motivation, as well as holding them responsible and accountable for output and outcomes of their actions so that they will contribute to their competence and satisfaction. Hence shared vision and shared mental models guide local decision makers [63]. In the case we view a computer system as a local decision maker than we have to make sure that the local decision maker acts according to the shared vision. Needless to say that (normative) control systems need to be in place to guide local decision makers and to make sure that local decision makers act within the boundaries fencing their decision power. Allocation of decision rights buttresses the notion of delegation. This (design) issue will be addressed later in this chapter. First we have to explore the notion of normative multi-agent systems.

4.2   Normative multi-agent systems: mechanistically viewed

Earlier we defined a normative multi-agent system as a system by means of mechanisms to represent, communicate, distribute, detect, create, modify, and enforce norms, and mechanisms to deliberate about norms and detect norm violation and fulfillment [2]. This definition takes the mechanism design point of view. In general a mechanism is a mathematical structure that models institutions through which for example economic activity is guided and coordinated. There are many kinds of these institutions for example law makers, administrators, managers of private companies like chief executive officers create institutions in order to achieve their desired goals [38]. The problem of mechanism design is: Given a class Θ of environments, an outcome space Z, and a goal function F, find a privacy preserving (i.e. a decentralized) mechanism $\pi = (M, \mu, h)$ that realizes F on Θ, where M is the message space, μ denotes the (group) equilibrium message correspondence $\mu : \Theta \mapsto M$ and h denotes the outcome function $h : M \mapsto Z$. The key insight of [37] was that information about the environment, facts that enable or constrain possibilities are distributed amongst agents. In the case an agent is not able to observe some aspect of the prevailing environment, than the agent does not have the information to guide his or her actions, unless the agent is communicated to by another agent who was able to observe. More specifically an agent is not able to observe the private information of another agent. Hence dispersion of private information amongst agents, known as information asymmetry, gives rise to specific incentive problems. By means of a verification scenario which separates the process of finding an equilibrium from recognizing an equilibrium it is possible to design incentive compatible mechanisms which occurs when the incentives that motivate the actions of individual participants are consistent with following the rules established by the group. Simply put in a verification scenario each agent reads the announced message by saying yes or no. The proposed outcome is judged acceptable if and only if the agents' responses are

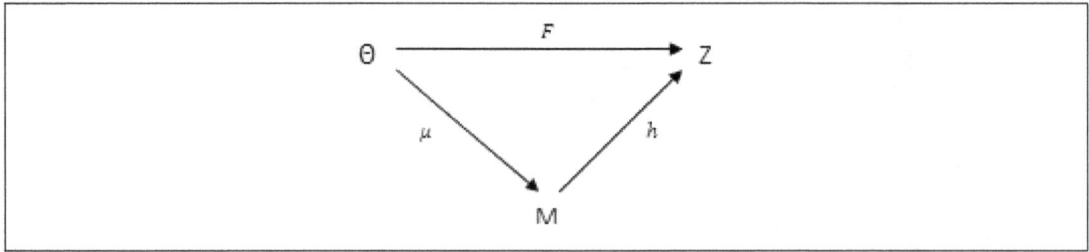

Figure 2: CommutingDiagram

affirmative. The message exchange process consists of three elements first a message space M, second a group equilibrium message correspondence μ, denoted $\mu : \Theta \mapsto M$ and third outcome function h, denoted as $h : M \mapsto Z$. A message space M consists of the messages available for communication. Messages may include formal written communication like contracts among buyers and sellers, accounting reports, production statistics, emails et cetera. Now it is easy to see that the group equilibrium message correspondence μ associates with each environment, θ, set of messages $\mu(\theta)$, that are equilibrium messages for all the agents. Assume that the messages were proposed actions, than $\mu(\theta)$ consists of all proposals to which each agent would agree in θ. The outcome function h translates messages into outcomes. As we have seen a mechanism π can be defined as an triple $\pi = (M, \mu, h)$. In the case π operates in a environment θ than the result is outcome $h(\mu(\theta))$ in Z. In the given space Θ, for all environments, the mechanism π leads to the desired result by the agent in that particular environment, than we say that π realizes F on Θ. In short π realizes F if for all θ in $\theta, h(\mu(\theta)) = F(\theta)$. Actually the equilibrium message μ represents the behavior of the agents. The concept can be represented in a commuting diagram as shown in figure 2

4.3 Effectiveness

The performance of a mechanism just described depends on elements that constrain the situation, like technological possibilities or those elements that define (i.e. influence) preferences of an agent, that are not subject to control or influence of the designer of the mechanism. The totality of all these elements is coined as the environmental space. In our exposition Θ. Furthermore we know that it is the case that no one, including the designer knows the prevailing environment θ and thus does not know the group equilibrium message μ. Agents know only their own parameters, the designer knows the environmental space Θ and the goal function F, informally defined as the class of environments for which the mechanism is to be designed for and the criterion of desirability. Remember the goal function F, reflects the agents criteria for evaluating the outcomes for example efficiency criteria, fairness criteria and so on. Indeed the mechanism provides in a logic in which ethical and moral criteria can be evaluated. These criteria are widely known to be effectiveness criteria [59]. As we will see later the notion of effectiveness plays a major role in designing all sorts of control and monitoring systems ensuring that goals of the designer are met. There are two crucial aspects we did not pay attention to namely the notion of the game form solutions concept and the revelation mechanism.

4.4   Game form mechanism

More precise the mechanism can be formulated in a game theoretical normal form. The game is defined by the agent's strategies, $S^1 ...S^N$ and their pay-off functions $\Psi^1...\Psi^N$. The joint strategy space is the Cartesian product of the agents strategies denoted as $S = S^1 \times ... \times S^N$. The pay-off function represents the utility of the agent when the joint strategy $s = (s^1...s^n)$ is used. The value h(s) refers to the outcome when the joint strategy s is used and the value of the pay-off function ,when s is used, is the value of the composite $\Psi^i(s) = \psi^i(h(s))$, where $\psi^i(s)$ denotes the utility of h. The game can be written as $G = G(S, h)$, so in the case the environment is θ, the utility function allows the agent to evaluate the pay-off from the joint strategy, when the outcome function is h. The solution concept and the specified message space induces privacy preserving group correspondences from the identified parameter space into the message space M, to be identified with the correspondences μ^i and μ. In the case N-tuple $(\mu^1...\mu^n)$ is an equilibrium, whatever type, the resulting messages in each environment θ, defines the correspondence Θ to M. Now we can easily see that once G implements a goal function F, then there is a mechanism π realizing F only in de case the correspondence μ makes the diagram commute (figure 2).

4.5   Incentive compatible and the revelation principle

Earlier we stated that an agent only knows his own characteristic, that is in the case of environment θ , agent i knows θ^i and his behavior depends only on the private information he has. We do agree on the fact that communication buttresses institutions like markets, organizations et cetera [33]. Opportunities for mutually hopefully beneficial transactions, social encounters et cetera cannot be found unless individuals share information about their preferences and endowments. The revelation principle states that, for many purposes, it is sufficient to consider only a special class of mechanisms, called incentive-compatible (direct or encoded) revelation mechanisms.

As we have seen the key idea is that each individual is asked to report his private information to a mediating mechanism. A direct-revelation mechanism is said to be 'incentive compatible' if, when each individual expects that the other persons (or agents in the sense of computer systems) will be honest and obedient to the mediating mechanism, then no individual (or agent) could ever expect to do better, given the information available to him, by reporting dishonestly to the mediating mechanism. So as a consequence the mechanism is incentive compatible if and only if honesty and obedience is in equilibrium of the resulting communication game. Hence for any equilibrium of any general communication mechanism, there exists an incentive-compatible (direct or encoded) revelation mechanism that is equivalent. This proposition is the revelation principle [51]. Hence the notion of equivalency (classes) determine under what conditions the commuting diagram commutes. Observe that it is also assumed that the mediating mechanism also needs to be honest and obedient. These values seem to guide actions of all participants in the game and therefore should be accounted for in the design of the mechanism that realizes the game. Participants in the game need to trust one another so each participant must be able to verify the proposed outcome often coined as transparency. In the case agents (human or computer systems) participate in a game than an agent is as well a principal as an agent. Stated otherwise their relationship is by nature reciprocal. Technically from a machine (that is a computational) point of view we recognize the notion of symmetric bi-directionality.

4.6   Incentives

In the case a principal delegates a task to an agent who has different objectives than delegating this task becomes problematic when the information about the agent is imperfect. Hereafter we will explicitly make a distinction between human agents and agents which are computer systems. Following [44] "If the human agent has a different objective function but no private information, the principal could propose a contract that perfectly controls the agent and induces the latterâĂŹs actions to be what he would like to do himself in a world without delegation". As a result incentive problems disappear. Alas conflicting objectives and decentralized information are thus the two basic ingredients of incentive theory. We will argue that even though objectives do not conflict or information is centralized that incentive problems still can occur. Think of fraud, criminal organizations, bribery, market misconduct, CEO compensation, slavery, environmental pollution et cetera. Here we enter the realm of norms and normative behavior. Three types of problems might occur in the case the human agent with private knowledge. First we have moral hazard or hidden action issues. Secondly we have adverse selection of hidden knowledge and thirdly the case of non verifiability. Non verifiability relates to the issue of sharing ex-post the same information but that no third party or no court of law can observe this information.

4.7 Some observations

When μ represents the actual behavior than this mechanism is compatible with the social definition of normative multi agent systems. In [2] a normative multi agent system is defined as "a multi agent system governed by restrictions on patterns of behavior of the agents in the system that are actively or passively transmitted and have a social function and impact". Patterns are represented as actions to be preformed, dictating what actions are permitted, empowered, prohibited or obligatory under a set of conditions and the specified effects when compliant and the consequences being not compliant with the set of conditions i.e. the norms. Hence the group equilibrium message correspondence μ associates with each environment, θ, set of messages $\mu(\theta)$, that are equilibrium messages for all the agents. In the case the messages contain proposed actions, than $\mu(\theta)$ consists of all proposals to which each agent would agree in θ. So the equilibrium message μ actually represents the behavior of the agents. The introduction of a verification scenario in the mechanism which separates the process of finding an equilibrium from recognizing an equilibrium makes it possible to design incentive compatible mechanisms which occurs when the incentives that motivate the actions of individual participants are consistent with following the rules established by the group. This is exactly what we are trying i.e. aiming to achieve following the rules of the group. A verification scenario warrants that each agent reads the announced message by saying yes of no. The proposed outcome is judged acceptable if and only if the agents' responses are affirmative. The goal function F reflects the agents criteria for evaluating outcomes, attributes of the outcome concern objectives like efficiency, fairness, effectiveness, compliance, reliability, trust whatever objectives formulated by the designer of the mechanism. It should be noted and emphasized that the equilibrium is founded in the logic of the price mechanism i.e. demand and supply mechanisms. In the case of the model proposed by Hurwicz the application was demonstrated using Walrasian tantonnement mechanism with continue price functions. The price mechanism in general has sound mathematical properties we will explore in the next chapter. The mathematical properties can be revealed using commuting diagrams. Commuting diagrams are a simple means to display some objects, linked together with arrows representing in our case functions. Commutativity means that the two paths depicted in figure 2 amount to the same thing; any two paths of arrows in the diagram that start at the same object and ends at the same object compose to give the same overall function. Instead of arrows the term morphisms is also used. So arrows are all set functions which in each appropriate case satisfy conditions relating to that structure[29]. We are interested in the way the arrows behave and the commonality they all have.

5   Relationships and values

The price mechanism ensures that a market price of a good and or service accurately summarizes the vast array of information held by market participants [6]. In the case prices depict i.e. fully reflect all available information than we say that the market is efficient[15]. Critics argue that markets are imperfect due to a range of all sorts of cognitive biases. But it needs no elaboration that the information aggregation characteristic of the pricing system buttresses many theories about the communicative function of prices and the decisions participants in the marketplace make to exploit business opportunities in the creation of value. It is the notion of value in this respect what is of interest for our purpose designing systems, particularly normative multi agent systems. Prices in economic theory reflect the value of an exchange in the marketplace such as buying and selling transactions. In early social theory the exchange mechanism is also used in analyzing social and anthropological mechanisms. Simmel uses the economic concept of value and argues that we should make a distinction in the exchange of value and the value exchange [64]. His first observation was that economic value is not just value in general but a definite sum of value, resulting from the commensuration of two intensities of demand to be exact the exchange of sacrifice and gain. An exchange is not a by-product of the mutual valuation of objects but its source [4]. For our purposes it suffices to look at value as defined in sociology, economics and anthropology[30]. Sociological concept of value is merely a conception of what is a good, proper, or desirable way to behave. In economic sense value refers to the degree to which objects are desired i.e. wants as measured how much others are willing "give up" to "get" these objects. Linguistically value might be defined as a meaningful difference. Hence they are all refractions of the same thing. Indeed they have some things in common and they might even share some properties. In the next section we explore the exchange mechanism in more detail.

5.1   Exchange mechanism characteristic

To describe a social encounter in a restaurant or an economic transaction in the marketplace in most cases agent models are used. Assume that we have an agent A who is willing to sell wool in some quantity at some price, given some quality standard. There is another agent B who is willing to sell cloth in some quantity at some price. Indeed there are different standards of quality in cloth, so the cloth for sale has some quality standard. Suppose agent A wants to buy cloth, and agent B wants to buy wool. The key question is then: How much is either agent willing to "give up" (i.e. to sacrifice) to "get" (i.e. to gain) the wool respectively the cloth? Typically this formulation captures precisely what is exchanged. It does not say anything how or when exchanges will actually occur. Exchanges are by definition reciprocal in nature and come in a large variety of what we coin as means like signed contracts, shaking hands et cetera. For example signing a contract by both parties is performative in nature; by the act of signing, we communicate that the exchange is done. Hence a signed contract affords exchanging. An affordance establishes the relationship between an object or an environment and an organism here a (human) agent through a stimulus to perform an action. In our example the stimulus is the signed contract and the detectable change in the external environment. We assume that the agent is sensitive and therefore able to respond to external (or internal) stimuli. This presumption is known as sensitivity. We have to realize that affordances are very special in nature. Following Gibson affordances of an environment are in a sense objective, real, physical unlike values and meanings, which are subjective, phenomenal and mental. But if we look closer than we must assert that affordances are neither an objective property nor a subjective property; hence they are both objective and subjective. It is equally a fact of an environment and a fact of behavior. An affordance points both ways, to the environment and to the agent (observer) [26]. Agents occupy niches of the environment, where we define a niche as a collection of affordances. Hence often we use terms like habitat or social space of an agent. Affordances can be measured with scales and standard units used for example in physics but they are as we have seen not just physical properties for they have unity relative to the posture and behavior of an agent. In our exposition we are interested in the unity relative to (moral) values. It seems that there are two interrelated notions of cognition at work: (1) experimental cognition and (2) reflective cognition. Experimental cognition leads to a state in which a (human) agent perceives and react to events around us. The reflective mode of cognition is about compulsion, contrast, thought and decision making. The difference lies in the technical details of the information structures of our brain; experimental cognition involves data-driven processing, where reflective cognition involves planning. Observe that via reflective mode of cognition we train i.e. learn to become an expert whose skill is that of experimental cognition[55]. Here we enter the realm of preferences, utility and the notion of bounded rationality. We will explore these notions later on in this chapter. First we return to our example i.e. problem introduced in this chapter.

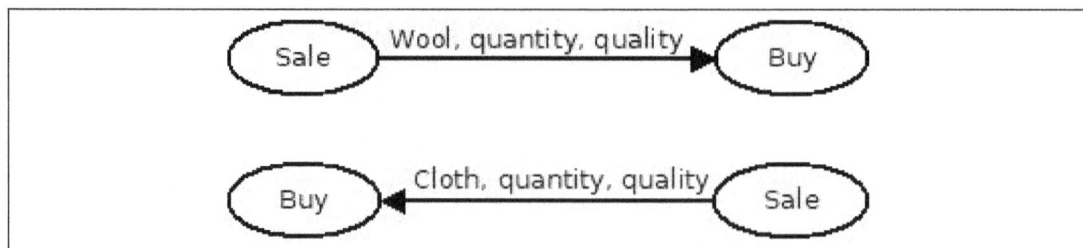

Figure 3: Barter Exchange

5.2 The exchange cycle: value exchange - exchange of values

Remember the situation in which an agent A who is willing to sell wool in some quantity at some price, given some quality standard. We have another agent B who is willing to sell cloth in some quantity at some price and that there were different standards of quality in cloth, so the cloth for sale has some quality standard. Now agent A wants to buy cloth, and agent B wants to buy wool. Our key question was: How much is either agent willing to "give up" (i.e. to sacrifice) to "get" (i.e. to gain) the wool respectively the cloth? The situation can be depicted graphically as follows.

We think it needs no elaboration that sacrifices and gains balance when equilibrium is reached. Informally our key question can be rephrased algebraically as:

$$Q^S_{W_{(q)}} : Q^B_{C_{(q)}} = Q^S_{C_{(q)}} : Q^B_{W_{(q)}} \tag{1}$$

Where

- The quantity of some object O, is denoted as Q

- The seller, denoted as superscript S of some object O

- The buyer, denoted as superscript B of some object O

- The object wool, denoted as subscript for some object O here Wool, denoted as subscript W of object Wool

- The object cloth, denoted as subscript for some object O here Cloth, denoted as subscript C of object Cloth

- The quality, denoted as subscript of a object (q)

Now we have to extend the model with the notion of measurement and valuation. In general a quantity is a property of a phenomenon, body or substance. Basically quantities are organized in a system of dimensions - SI. There are so-called base quantities like length and time for example, and quantities that are derived from these base quantities. Hence each base quantity has its own dimension which property has a unique magnitude that can be expressed as a number and a reference. Observe that each derived quantity's dimension follows from the derivation itself [32]. Rearranging our equation we get:

$$\frac{Q^S_{W_{(q)}}}{Q^B_{C_{(q)}}} = \frac{Q^S_{C_{(q)}}}{Q^B_{W_{(q)}}} \tag{2}$$

Now we extend our equation with the notion of measurement for the quantity of object O:

$$\frac{Q^S_{W_{(q)}}}{Q^B_{C_{(q)}}} \cdot \frac{Q^{st}_W Q^m_W}{Q^{st}_C Q^m_C} = \frac{Q^S_{C_{(q)}}}{Q^B_{W_{(q)}}} \cdot \frac{Q^{st}_C Q^m_C}{Q^{st}_W Q^m_W} \tag{3}$$

Where the quantity of the object O is measured in some standard unit expressed as a number and a reference denoted as superscript st and superscript m, the dimension quality denoted as (q) of object, the dimension absolute frequency as a number of objects. Standard units expressed as a number and a reference $Q^{st}_O Q^m_O$ in (3) can be denoted as $U_{(O_q)}^S$ for the sell side and $U_{(O_q)}^B$ for the buy side, where U denotes the standard unit expressed as a number and a reference. The quantity of the object O is measured in some standard unit U and the measurement is expressed as a product $Q \cdot U$, the dimension quality denoted as q of object, the dimension absolute frequency as a number of objects. We get:

$$\frac{Q^S_{W_{(q)}}}{Q^B_{C_{(q)}}} \cdot \frac{U^S_{W_q}}{U^B_{C_q}} = \frac{Q^S_{C_{(q)}}}{Q^B_{W_{(q)}}} \cdot \frac{U^S_{C_{(q)}}}{U^B_{W_{(q)}}} \tag{4}$$

Analogue to the definitions of equation (4) we can write (5) with the notion of measurement defined as U for the quality of objects O as:

$$\frac{Q^S_{W_{(q)}}}{Q^B_{C_{(q)}}} \cdot \frac{U^S_{W_q}}{U^B_{C_q}} \cdot \frac{U^S_W}{U^B_C} = \frac{Q^S_{C_{(q)}}}{Q^B_{W_{(q)}}} \cdot \frac{U^S_{C_{(q)}}}{U^B_{W_{(q)}}} \cdot \frac{U^S_C}{U^B_W} \tag{5}$$

It is quite easy where the money part comes in, just multiply the equations with a unit of measurement for money. Let v is the dimension of currency, denoted as v for the money unit. Observe that in our equation each base quantity has its own dimension which property has a unique magnitude that can be expressed as a number and a reference. Hence the unit v is easy extensible to a multitude of currencies. For our purposes we leave this subject to a rest. In the case we define the unit of measurement for money as we get:

$$\frac{Q_{W_{(q)}}^S}{Q_{C_{(q)}}^B} \cdot \frac{U_{W_q}^S}{U_{C_q}^B} \cdot \frac{U_W^S}{U_C^B} \cdot \frac{U_v}{U_v} = \frac{Q_{C_{(q)}}^S}{Q_{W_{(q)}}^B} \cdot \frac{U_{C_{(q)}}^S}{U_{W_{(q)}}^B} \cdot \frac{U_C^S}{U_W^B} \cdot \frac{U_v}{U_v} \tag{6}$$

It is straightforward to see when we multiply all variables i.e. factors that we only have a money measure and that all information is encapsulated in the money measure. When markets are efficient in the way Fama formulated than it is very useful to use market based measures for evaluation procedures[15, 58, 57]). As we mentioned earlier this line of reasoning met some critiques from theorists and experimentalists that the behavioral assumptions underlying the information aggregation characteristic are flawed. In general the critiques concentrate on the rationality assumptions and that one should look for evidence about what humans actually do [8]. As for economists, sociologist, anthropologists and psychologist equation (5) is quite interesting. Economists are interested in the price behavior and market conditions. For example market regulation issues like monopolistic behavior and transaction cost economics coined as market and organizational failure theories [71]. Basically Williamson argues that bounded rationality characteristics combined with opportunistic behavior of agents are major concerns classical economic organizational theories overlooked. Economists tend to integrate psychology to economic theory to explain (economic) behavior. Aspects are altruism, happiness, pro-social behavior, the helping hand et cetera [19]. Very interesting is some older work of Mauss [47]. He studied the actual act of exchange of gifts and rendering of services, and the reciprocating or return of these gifts and services. Although there was no economic system as we know it, Mauss argues that the society he studied can be described by the catalogue of transfers that map all the obligations between its members. The cycling gift system is the society. If we look at equation (5) than we could say that the left part is the weighing function of agent A and the right part is the weighing function of agent B. The weighing function becomes hard to decipher in the case the weighing function is not linearizable. As a consequence we cannot verify whether the calculations are properly conducted. Hence outcomes become unpredictable. Theorists introduce therefore utility functions and conditions that are assumed like preference ordering characteristics and monotonicity. Hence if we are able to measure the goods or services in the correct unit of measurements and the only uncertainty is the outcome of the quality evaluation of agent A and agent B towards the objects sold and bought, than in the case agent A and agent B come to an agreement we know that the equations (7) and (8) must hold:

$$P_B(Q^B_{W_{(q)}} \cdot U^B_{W_q} \cdot U^B_W) > P_A(Q^S_{W_{(q)}} \cdot U^S_{W_{(q)}} \cdot U^S_W) \tag{7}$$

$$\bigwedge$$

$$P_A(Q^B_{C_{(q)}} \cdot U^B_{C_q} \cdot U^B_C) > P_B(Q^S_{C_{(q)}} \cdot U^S_{C_{(q)}} \cdot U^S_C) \tag{8}$$

P_A en P_B denote the preference function outcomes of the objects bought.

5.3 Design characteristics

The structure of the preference function of the agents is what we actually need to elaborate upon in the case norms are modeled for society and the design of the normative multi agent system is value sensitive by design. As we have seen affordances can be measured with scales and standard units used for example in physics but they are not just physical properties for they have unity relative to the posture and behavior of an agent. We are interested in the unity relative to (moral) values. Our equations (7) and (8) formulate precisely the decision rule i.e. the procedure to reason about whether the monitored systems actually behaves in an ethical i.e. moral fashion, actually this is what a normative multi agent system does and foremost the designer can actually formulate which or whenever design choices have to be made, why the choices are inevitable and what consequences these choices actually have for the design and effectiveness of the artifact. Hence the model equation (1) depicts the most elementary mechanism of any exchange relationship. Using the elementary units organized in a system of dimensions we actually enrich the model so we can ensure that no information ever gets lost. Indeed it ensures the minimum informational requirements warranting consistency of the data processing facilities like software, algorithms, communication, hardware, networks, and search and database technologies. Consistency is paramount and buttresses the notion of data quality i.e. data integrity. For example ACID (Atomicity, Consistency, Isolation, Durability) is a set of properties that guarantee that database transactions are processed concurrently. Hence if we design a distributed environment based on web services the key design question is: "How to ensure the ACID principles in transaction processing using web services?"[28]. The same design question can be formulated for workflow systems such as inter and intra- organizational workflows, contracts nets, value nets et cetera [9]. Observe that the first design question provides in a mechanism to ensure the second design question. So we only have to concentrate on the special requirements on the process level ensuring external integrity of the information processing function of the artifact. Observe that the model depicted as equation (4) and the decision rules equation (7) and (8) realizes the mechanism π described in chapter

4, under the conditions that elementary units organized in a system of dimensions are used in the model. We observed that when μ represents the actual behavior that this mechanism is compatible with the social definition of normative multi agent systems. The introduction of a verification scenario in the mechanism which separates the process of finding an equilibrium from recognizing an equilibrium makes it possible to design incentive compatible mechanisms which occurs when the incentives that motivate the actions of individual participants are consistent with following the rules established by the group. There is one (big) difference: we did not use utility functions, but instead we formulated a preference ordering derived from the systems of dimensions. The valuation itself is an empirical question and should be treated as such because we need models that have high predictive value. Otherwise the designed mechanism like normative multi agent systems fails to realize the goal function of the design. We separated the preference and ordering conditions from the object and the subject. Therefore human peculiarities in decision making can be studied in isolation and in combination with the environment. Hence the behavior is influenced by, depends on the environment. We think that the human preference orderings behave on a continuum where complexity and uncertainty plays an important role. Notions of this adaptive toolbox describe mechanisms to model adaptive behavior of agents in the environment they "see" [27].

6   Principles, architectures and state transition systems

Like we stated earlier a model is always a result and the start of a design process. "A (design) process is an abductive sensemaking process, a step of adopting a hypothesis as being suggested by the facts ... a form of inference, albeit inference of "best guesses" leaps [..]. A logic of what might be. It is not entirely accurate,... it is the argument to the best explanation, the hypothesis that makes the most sense given observed phenomena or data and based on prior knowledge" [43]. This is precisely the function of our model depicted as equation (4) and the decision rules equation (7) and (8) realizes the mechanism π described in chapter 4. We return to our Nike example. Suppose that agent A is Nike, and agent B is one of the suppliers in the direct supply chain. Nike wants to be sure that agent B is compliant with applicable laws and regulations ensuring that slavery is rooted out from the supply chain. In the case agent B delivers goods manufactured under conditions of slavery than we would expect that equation (8) fails and thus is not true. What information does agent A (Nike) need to make this assertion? But we have another issue to address simultaneously: how can we be sure that agent A (Nike) will be truthful in their actions and communications. Stated otherwise how to distinguish moral hazard, from adverse selection and non-verifiability problems? We will have to formulate constitutive rules and regulative rules, grounded in our belief, our attitude, et cetera [2]. This process is iterative in nature, and the model facilitates current understanding among participants in the direct supply chain and its stakeholders and supervisors. Hence all aspects identified in chapter 2 will be addressed and henceforth all types of ethics will be addressed to reason about the purpose of the normative agent system fuelling the question and answering how we can design a mechanism that actually realizes the goal function of Nike and the goal function of in this case the legislator. We have identified seven key questions which have to address in modeling norms. These were:

	Questions	Sets
1.	What is the believe of the state of California with reference to slavery and human trafficking?	$\mapsto$B
2.	Are the values expressed by extricating slavery and human trafficking from direct supply chains subsumed in other values?	$\mapsto$F
3.	Are there several distinct values expressing extricating slavery and human trafficking?	$\mapsto$P,M
4.	Who are the value bearers in the supply chain?	$\mapsto$N
5.	How are the decision rights dispersed in the supply chain, who is responsible and accountable ?	$\mapsto$DRAP,N
6.	What rules, standards, regulations, rewards and punishment are established preserving moral values in the direct supply chains?	$\mapsto$N
7.	If applicable is there a representation expressing the moral value in communication processes?	$\mapsto$S

A design principle is a normative principle on the design of the artifact as such ,it is a declarative statement that normatively restricts design freedom [12, 31]. So by answering the questions summed up above we elicit the normative principles on the design of the artifact. Requirements defined as a required property of an artifact also limits design freedom. Indeed requirements state what properties an artifact should have from the perspective of the goals of stakeholders i.e. institutions, legislators, supervisors, society et cetera. Goals motivate why requirements are imposed on the design. The first three questions address design principles used to express, i.e. buttresses policies to ensure that the design of the artifact meets the aforementioned requirement defined as a property that the artifact should have realizing the goal function of the stakeholder(s. The questions address the What and Why of the design. The questions 4 and 5 address the notion of moral agency. Indeed **who** is responsible as accountable? By answering question 6 we address **how** norms as moral values are enforced where question 7 addresses the informational (infra)structure like information processing, communications and storage of data i.e. how communication processes enables the interaction among agents (human and machines). Observe that the identified questions address governance and management perspectives [40, 56]. The first thee questions cover the goal setting processes and objective setting i.e. the governance system, while questions 4 and 5 cover the management system. Question 6 covers the process and control dimension where question 7 covers the information systems and infrastructure. Models in general are to be understood as purposeful abstractions i.e. representations of (some) reality. Usage of models is to represent systems; actually the model can be regarded as a system in itself [3]. Models are analogous to Janus structures representations with an engineering side facing the real world and an abstract side facing theories[67]. It is possible even most likely that the model does not fit the empirical data, just because the theory was not appropriate so the theory i.e. our belief has to be revised. The revision process is actually a meta level technique for examining the axioms upon which the theory was "founded". By altering the axioms or postulates new theories are formulated that hopefully forms a better match with the facts. In design practice it is simpler to formulate theories in first order logics and use explicitly meta reasoning about axioms and postulates. Indeed we are interested in mechanisms that realize goal functions. This notion is known as the AGM axioms for theory revision[1]. In the case our model does not realize the goal function expressed as a belief than we examine whether the pre-conditions i.e. the axioms and postulates buttressing the model are appropriate. Axioms and postulates are directed internally representing the intentional internal point of view. For example when we return to our example of Nike question 1 covers the belief, where questions 2 and 3 cover the intentional point of view, as we have seen being the central structure of an experience directed towards an object by virtue of its content or meaning which represents the object [5]. This feature characteristic will be important when we ask ourselves whether a computer system i.e. an artifact can be a moral agent. Next we explore the notion of architecture as a model representing a system.

6.1 Architecture as a model representing a system

The IEEE defines an architecture as "the fundamental organization of a system embodied in its components, their relationships to each other, and to the environment, and the principles guiding its design and evolution" [39]. Basically every information system is an assembly of 5 basic components known as the von Neuman Architecture [52]. We have input and output devices, a CPU containing a control unit and ALU and a (internal) memory unit. Computationally a representation defined as a pattern of symbols that stand for values are coined as data and when implemented by a computer system an algorithm controls the representation as input and the representation as output so the algorithm controls the transformation of data representations[11]. Theoretically there are several models in which the actual behavior of a discrete system can be described. All these models can be described as state transition systems. Formally computation can be studied by means of a state transition systems defined as a pair $(S, \rightarrow)$ where S is a set of states and $\rightarrow$ is a set of transitions; the state p to state q denoted as $(p, q) \in \rightarrow$ we write $p \rightarrow q$. It is easy to label a transition. Labels can mean anything, like expected input conditions, actions to perform during the transition, conditions that must be true before triggering a transition. The state transition system with label's is a tuple $(S, \Lambda, \rightarrow)$ where S is a set of states, $\rightarrow$ is a set of state transitions and Λ is a set of labels; the state p to q with label α. In the case we are not familiar with the semantics of the label or simply put the semantics of the labels are not known to us, a labeled deductive system (LDS) as a means to be able to reason properly about the representation in the system seems to be a necessary first step to translate the logic of a state transition system with labels via a LDS to classical logics. Hence an LDS is family of logics[24]. Observe that the AGM axioms for theory revision can be formulated as an LDS and translated to classical logics. Architecture is the normative restriction of design freedom [12, 31].

7 Ethical sensitive design

7.1 Decision right allocation procedure (DRAP) and the verification mechanism

We defined human agency as the capacity to make choices and entails the claim that humans do in fact make decisions and enact them on the world. The key design problem in design processes is how to address reflexivity. In chapter 2 we elaborated on the notion of reflexivity as being the mechanism referring to an act of self reference recognizing forces or pressure within the environment and his or her place in the social structure. Agents with a low level of reflexivity are said that the environment shapes the individual norms, tastes, wants et cetera. In the case agents with a high level of reflexivity shape for example their own norms and tastes. Reflexivity addresses autonomy and thus autonomous action of an agent. This is exactly where principles in general and moral principles as moral values come into play, namely principles restrict autonomy of an agent. We like to defend that reflexivity refers to rule-based ethics versus virtue-based ethics. Rule-based ethics is governed by concepts like acts, moral rules and moral principles and virtue ethics is governed by moral dispositions, emotions, states of character and the flourishing of human beings. In virtue ethics morality is directly linked i.e. intimately linked to the person who acts, to his or her character and situation. This notion demarcates social space among agents. Agents occupy niches of the environment seen as the world, where we define a niche as a collection of affordances, often coined as habitat or social space of an agent. An environment is in a sense objective, real, physical unlike values and meanings, which are subjective, phenomenal and mental. Affordances are neither an objective property nor a subjective property; they are both objective and subjective. It is equally a fact of an environment and a fact of behavior. An affordance points both ways, to the environment and to the agent (observer) [26]. We cannot say in advance that reflexivity behaves on a continuum from rule-based ethics to virtue-based ethics and vise versa. So in the design process we have to make provisions to decide upon how consensus can be reached among agents. Here we must decide what type of rules we adopt to verify whether consensus is reached. Observe consensus in a design setting addresses the shared values among agents. Each agent will make their private considerations whether to agree or not. Indeed it is possible to reconsider the earlier made choices. Whether this type of rule is accepted is a fundamental design question addressing agency. In some cases it is impossible to decide upfront how a mechanism is to be designed, build and implemented. This means that deciding upon a verification procedure recognizing whether consensus is reached among agents must have provisions for reconsidering earlier made choices. Indeed all seven questions categorized in what, why, who en how are permanently defied by agents and therefore monitoring warrants the soundness of moral choices being made by agents. This is precisely what a normative multi agent system does; monitoring warranting the soundness of moral choices, recognizing whether agents are not compliant, and recognizing whether the designed moral system itself does not have possible negative side effects subverting the moral system actually desired [7, 35, 48].

7.2 Creating a vision from first principles

By answering and discussing the seven questions that guide action in designing a normative multi agent system our aim is to create a shared vision and thus shared mental models that guide local decision makers i.e. the agents.

A vision contains i.e. envisions the outcome of the deliberation process discussing and answering the seven questions in a coherent, consistent and sequacious way. Our design question is diagnostic in nature and the reasoning style is abductive.

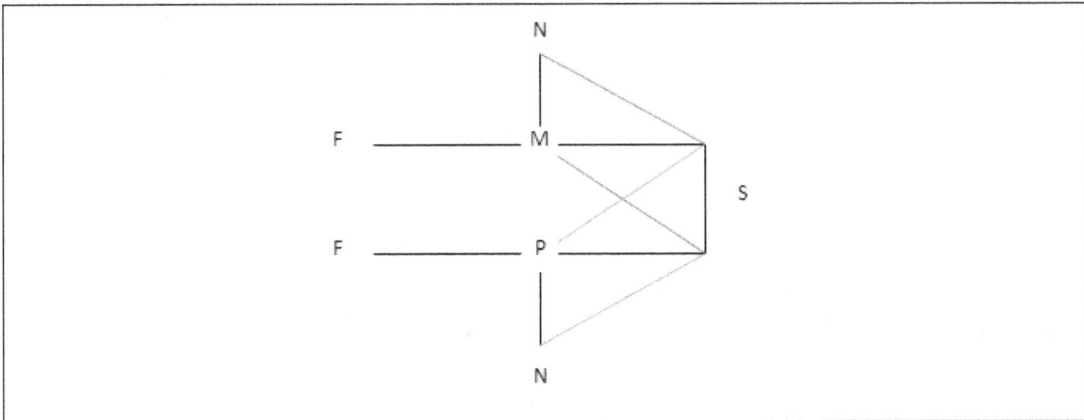

Figure 4: Value pluralism

Traditional the diagnostic problem is framed in situations where an observation of the system's behavior is functioning abnormal or even fails to function at all. The issue is than to determine those components, objects et cetera of the system that will explain the difference between observed behavior and the desired correct behavior [62]. To solve the aforementioned diagnostic problem from first principles only the information of the system description is available together with the observation of the actual behavior. Reiter builds on the work of [41] and provides in a theoretical foundation for diagnosis from first principles. For representation purposes Reiter choose first-order logic for representing systems. As he observed and demonstrated many different logics lead to the same theory of diagnosis. Hence more abstractly Reiter's theory can be formulated as a LDS and then translated into classical logic here first-order logic. In our situation there is one major difference and that is we cannot observe actual system behavior just because the system has to be designed yet. We do have a shared expectation about the expected behavior and we aim that the system after being build and implemented shows in practice the shared expected behavior. Indeed we have to consider that there is a possibility that the actual behavior after having the system built and implemented can actually differ from the expected outcome and we will need safeguards upfront to consider in designing the system. We coined this requirement incentive-compatible (direct or encoded) revelation mechanisms. It needs no elaboration that the design problem and the diagnostic problems both share the same mechanisms and principles. This is easy to see in figure 4 [41].

The main objective is to design a system that minimizes the expected structural discrepancy between the model and the artifact realizing the goal function. For ex-

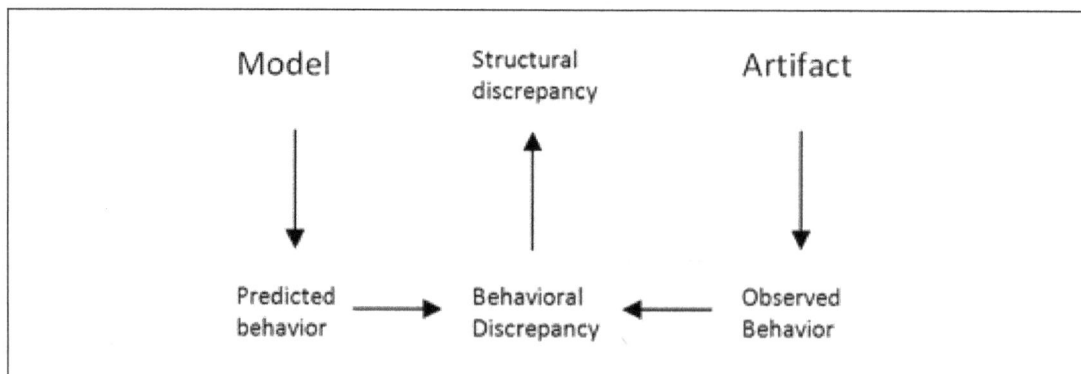

Figure 5: Diagnostic cycle

ample: if the moral values are not to be debated than a model based on the axioms of such a belief decided to be foundational and strict normative in the deontological sense than the normative multi agent system is to be designed to monitor the behavioral discrepancies between predicted that is normative behavior versus observed normative behavior; strict rules should be enforced upon the agents who are responsible as accountable on merit grounds, whether moral values are in the plural or monist like. Indeed the verification procedure applied by the normative multi agent system communicates the outcome of the verification procedure analogous to the group correspondence message π. In the case the equilibrium is not recognized than the normative multi agent system has to inform the agent whose action is not compliant to the applicable rule so corrective action can be taken or the agent is to be punished by some rule. Punishment can be a blaming and shaming mechanism, dissipation from the group, or group activities, imposition of individual fines, restrict autonomy et cetera. In the case the equilibrium is recognized than the group correspondence message π actually reflects the behavior of the agents so behavioral discrepancy is not observed and we may infer that the mechanism realizes the goal function of the normative system. Indeed analogous to the punishment ruling we can actually reward the agents for being compliant. Rewarding agents can either be financial, augment autonomy, pat on the back, more privileges et cetera. In this design the value bearer is the agent i.e. the individual. Indeed if we assume that the reflexivity of the agent is low than we might expect that the agent will adapt to his, hers, its environment. If the reflexivity of the agent is high than the key question is whether the agent is willing to comply or subverts the system by lying, cheating or neglects actions to perform and so on. As we have seen the design process provisions in a mechanism under what conditions consensus is reached and maintained. Suppose we have a majority rule, than we cannot rule out that an autonomous agent does not agree to the full extend what has been decided. In this latter case we cannot rule out by design possible behavioral discrepancy of the agent. There are two options for the designer(s). The first option is to take another close(r) look at the actual axioms, presumptions buttressing the model as in our example we started with in the first place. The second option is to introduce more rules and enforce harder. The designers have to make a choice: "which path to follow?" If it is possible to reconsider the earlier made choices than the revision process will be commenced. Whether this type of rule is accepted is a fundamental design question addressing moral agency.

References

[1] C. Alchourron, P. Gardenfors, and D. Makinson. On the logic of theory change: partial meet contraction and revision functions. *Journal of Symbolic Logic*, 50(2):510–530, 1985.

[2] G. Andrigetto, G Governatori, P. Noriega, and L.W.N. van der Torre. *Normative Multi-Agent Systems*, volume 4 of *Dagstuhl follow-Ups*. Schloss Dagstuhl, Saarbrucken / Waadern, first edition, 2013.

[3] L. Apostel. Towards a formal study of models in the non-formal sciences. *Synthese*, (12):125–161, 1960.

[4] A. Appadurai. *Introduction: commodities and the politics of value*, book section 1. Cambridge University Press, 1986.

[5] M. Ashmore. *The reflexive thesis*. University of Chigago Press, 1989.

[6] A.E. Atakan and M. Ekmekci. Auctions, actions, and the failure of information aggregation. *American Economic Review*, 104(7):2014–2048, 2014.

[7] D.V. Budescu and M. Bruderman. The relationship between illusion of control and desirability bias. *Journal of Behavioral Decision Making*, 8:109–125, 1995.

[8] C. Camerer. *Behavioral game theory: experiments in strategic interaction*. Princeton University Press, 2003.

[9] R.M.J. Christiaanse, P. Griffioen, and J. Hulstijn. Reliability of electronic evidence: an application for model-based auditing. In *Artificial Intelligence and Law*. AAAI - ACM SIGART, 2015.

[10] H. Coppens, T. Geel van. Urban architectural design and scientific research: how to save an arranged marriage, 2013.

[11] P.J. Denning and C.H. Martell. *Great Principles of Computing*. Massachusetts Institute of technology, 1st edition, 2015.

[12] J.L.G. Dietz. *Architecture - buiding strategy into design*. Academic service / SDU, The Haque, 2008.

[13] ECL. 2015:ecli:nl:rmmne:2015:3262.

[14] ECL. Ecli:nl:phr:2016:883.

[15] E.F. Fama and M.H. Miller. *The theory of finance*. New York, 1972.

[16] Luciano Floridi. Open problems in the philosophy of information. *Metaphilosophy*, 35(4):29, 2004.

[17] Luciano Floridi. *Foundations of Information Ethics*. The Handbook of Information and Computer Ethics. John Wiley & SOns, New Jersey, 2008.

[18] R. Foque. Research in design science. *ADSC*, 10-11, 2003.

[19] B.S. Frey and A. Stutzer. *Economics ans Psychology*. MIT Press, London, 2007.

[20] Friedman. *Human values and the design of computer technology*. Cambridge University Press, 1997.

[21] B. Friedman, P.H. Kahn, and A. Borning. Value sensitive design: Theory and methods. Report, Universtity of Washington, 2002.

[22] B. Friedman, P.H. Kahn, and A. Borning. *Value Sensitive Design and Information Systems*. John Wiley & Sons Inc., New Yersey, 2008.

[23] Kahn Friedman. *Human values, ethics and design.*, page 177âĂŞ1201. Lawrence Erlbaum, Mahwah, 2003.

[24] D.M. Gabbay. *Labelled Deductive Systems*, volume 1. Oxford University Press, Oxford, 1996.

[25] E.L. Gettier. Is justified true belief knowledge? *Analysis*, 23:(6):121–123, 1963.

[26] J.J. Gibson. *The ecological approach to visual perception*. Taylor Francis Group, New york, 1986.

[27] G. Gigerenzer. *Bounded rationality: the adapive toolbox*. MIT Press, Berlin, 2001.

[28] S. Gilbert and N. Lynch. Brewer's conjecture and the feasibility of consistent, available, partition-tolerant web services. *SIGACT News*, 33(2):51–59, 2002.

[29] R. Goldblatt. *Topoi, the categorial analysis of logic*. Dover, first edition, 2006.

[30] D. Graeber. *Toward an Anthropological Theory of Value*. Palgrave, New York, 2001.

[31] D. Greefhorst and E. Proper. *Architecture Principles*. Springer, 2011.

[32] P.R. Griffioen. Type inference for lineair algebra with units of measurements. Report, CWI Amsterdam, 2013.

[33] F. Hayek. The use of knowledge in society. *American Economic Review*, 35:519–530, 1945.

[34] A. Heylighen, H. Cavellin, and M Bianchin. Design in mind. *Design Issues*, 25(1), 2009.

[35] G. Hofstede. Management control of public and not-for-profit activities. *Accounting, Organizations and Society*, 6(3):193–211, 1981.

[36] Jeroen van den Hoven. *Handbook of Ethics, Values and Technological Design*. Springer, 2015.

[37] L. Hurwicz. *Optimality and informational efficiency in resource allocation processes*. Mathematical methods in social sciences. Stanford Universiy Press, 1960.

[38] L. Hurwicz and S. Reiter. *Designing economic mechanisms*. Cambridge University Press, New York, 2006.

[39] IEEE. Iso architecture, 2011.

[40] ISACA. Cobit 5, a business framework for governance and management of enterprise

it. Report, ISACA, 2012.

[41] J. Kleer de and B.C. Williams. Diagnosing multiple faults. *Artificial Intelligence*, 32:97–130, 1987.

[42] C. Kluckhohn. *The study of culture*. The Policy Sciences. Stanford University Press, 1951.

[43] J. Kolko. Abductive thinking and sensemaking: The drivers of design synthesis. *Design Issues*, 26(1):14, 2010.

[44] J-J. Laffont and D. Martimort. *Theory of Incentives: the principal - agent model.* Princeton University Press, Princeton, 2002.

[45] A Macintyre. Social structures and their threads to moral agency. *Philosophy*, 74(289):311 – 329, 1999.

[46] Mason, 2015.

[47] M. Mauss. *The Gift.* Presses Universitaires de France in Socialogie et Anthropologie, 1950.

[48] K.A. Merchant. The control function of management. *Sloan Management Review*, (Summer 1982):43–55, 1982.

[49] merriam webster. definition of a norm, 2016.

[50] Mustafa Emirbayer Mische and Ann. What is agency? *American Journal of Sociology*, 3(4):962âĂŞ1023, 1998.

[51] R Myerson. Optimal coordination mechanisms in generalized principal-agent problems. *Journal of Mathematical Economics*, 10:67–81, 1982.

[52] J. Neuman. First draft of a report on the edvac. Report, 1945.

[53] NIKE. Supply chain, 2016.

[54] NIKE. Transform manuafactuering, 2016.

[55] D.A. Norman. *Things that make us smart.* Basic Books, 1993.

[56] OECD. G20/oecd priciples of corporate governance. Report, OECD, 2015.

[57] W.G. Ouchi. An organisational failures framework. Report 461, Stanford, June, 1978 1978.

[58] W.G. Ouchi. Markets, bureaucraties, and clans. *Administrative Science Quarterly*, 25(1):129–141, 1980.

[59] L Paape. *Corporate Governance: the impact on the role, Position, and Scope of services of Internal Audit.* Thesis, 2007.

[60] R. Poivet. *Le realism esthetique.* Presses Universitaire de France, 2006.

[61] R. Poivet. Moral and epistemic virtues: A thomistic and analytical perspective. *International Journal for Philosophy*, 15(1):1–15, 2010.

[62] R. Reiter. The theory of diagnosis from first principles. *Artificial Intelligence*, 32:57–95, 1987.

[63] P.M. Senge. *The fifth discipline: the art and practise of learning organization.* Doubleday, New york, first edition, 1990.

[64] G Simmel. A chapter in the philosophy of value. *Journal of Sociology*, V(5):577–603,

1900.

[65] R. Simons. The role of management controlsystems in creating competitive advantage: new perspectives. *Accounting, Organizations and Society*, 15(no 1/2):16, 1990.

[66] R. Simons. *Levers of Control: How Managers Use Innovative Control Systems to Drive Strategic Renewal.* Harvard Business School Press., 1995.

[67] J.F. Sowa. *Knowlegde representation: logical, philosophical, and computational foundations.* Thomson, Pacific Groove, 2000.

[68] Stanford. Ethics-computer, 2015.

[69] P.F. Strawson. *Freedom and resentment and other Essays.* Routledge, 2008 edition, 1974.

[70] Jarvis Thomson. The trolly problem. *The Yale Law Journal*, 94(6):1395–1415, 1985.

[71] O.E. Williamson. *Markets and hierarchies, analysis and antitrust implications.* The Free Press, 1975.

[72] M. Woolridge. *Intelligent Agents*, volume 1, book section 1, page 609. MIT press, London, 2000.

[73] M. Woolridge. *An introduction to multiagent systems*, volume 1. John Wiley & Sons, Chichester, 2011.

 Received 5 September 2016

www.ingramcontent.com/pod-product-compliance
Lightning Source LLC
Chambersburg PA
CBHW080516090426
42734CB00015B/3072